AF807328

Palgrave Studies in Green Criminology

Series Editors
Angus Nurse, Nottingham Trent University, Nottingham, UK
Rob White, School of Social Sciences, University of Tasmania, Hobart, Australia
Melissa Jarrell, Department of Social, Cultural, and Justice Studies, University of Tennessee, Chattanooga, USA

Criminologists have increasingly become involved and interested in environmental issues to the extent that the term Green Criminology is now recognised as a distinct subgenre of criminology. Within this unique area of scholarly activity, researchers consider not just harms to the environment, but also the links between green crimes and other forms of crime, including organised crime's movement into the illegal trade in wildlife or the links between domestic animal abuse and spousal abuse and more serious forms of offending such as serial killing. This series will provide a forum for new works and new ideas in green criminology for both academics and practitioners working in the field, with two primary aims: to provide contemporary theoretical and practice-based analysis of green criminology and environmental issues relating to the development of and enforcement of environmental laws, environmental criminality, policy relating to environmental harms and harms committed against non-human animals and situating environmental harms within the context of wider social harms; and to explore and debate new contemporary issues in green criminology including ecological, environmental and species justice concerns and the better integration of a green criminological approach within mainstream criminal justice. The series will reflect the range and depth of high-quality research and scholarship in this burgeoning area, combining contributions from established scholars wishing to explore new topics and recent entrants who are breaking new ground.

Simon Sneddon

Food, Environmental Degradation and Injustices

palgrave
macmillan

Simon Sneddon
Department of Law
University of Northampton
Northampton, Northamptonshire, UK

ISSN 2946-269X ISSN 2946-2703 (electronic)
Palgrave Studies in Green Criminology
ISBN 978-3-031-87899-2 ISBN 978-3-031-87900-5 (eBook)
https://doi.org/10.1007/978-3-031-87900-5

© The Author(s), under exclusive license to Springer Nature Switzerland AG 2025

This work is subject to copyright. All rights are solely and exclusively licensed by the Publisher, whether the whole or part of the material is concerned, specifically the rights of translation, reprinting, reuse of illustrations, recitation, broadcasting, reproduction on microfilms or in any other physical way, and transmission or information storage and retrieval, electronic adaptation, computer software, or by similar or dissimilar methodology now known or hereafter developed.
The use of general descriptive names, registered names, trademarks, service marks, etc. in this publication does not imply, even in the absence of a specific statement, that such names are exempt from the relevant protective laws and regulations and therefore free for general use.
The publisher, the authors and the editors are safe to assume that the advice and information in this book are believed to be true and accurate at the date of publication. Neither the publisher nor the authors or the editors give a warranty, expressed or implied, with respect to the material contained herein or for any errors or omissions that may have been made. The publisher remains neutral with regard to jurisdictional claims in published maps and institutional affiliations.

Cover illustration: © Melisa Hasan

This Palgrave Macmillan imprint is published by the registered company Springer Nature Switzerland AG
The registered company address is: Gewerbestrasse 11, 6330 Cham, Switzerland

If disposing of this product, please recycle the paper.

For Katie

Competing Interests The author has no competing interests to declare that are relevant to the content of this manuscript.

Contents

CHAPTER 1

Introduction

Abstract This introductory chapter explores the relationship between food systems and the multifaceted injustices arising from environmental degradation. Drawing on Rob White's eco-justice framework, the text uses his three interconnected dimensions: environmental (human-centred impacts), ecological (ecosystem-centred impacts), and species based (focusing on non-human life). The book is positioned as an extended thought experiment, imagining the cessation of specific food commodities and examining the resulting ripple effects through ethical, economic, and ecological lenses. It adopts a neutral stance, critically evaluating industrial food systems and their role in perpetuating inequities, biodiversity loss, and environmental harm. Key concepts like environmental justice emphasise the disproportionate burdens on marginalised communities, while ecological justice underscores the intrinsic rights of ecosystems. Species justice interrogates the welfare and sentience of non-human life, especially in intensive farming contexts. The chapter outlines the book's methodology, examining the sustainability and justice implications of alternative food sources, including novel and lab-grown products. By imagining radical shifts in food systems, the chapter invites reflection on how global agriculture can be reimagined to prioritise equity, resilience, and sustainability.

© The Author(s), under exclusive license to Springer Nature Switzerland AG 2025

S. Sneddon, *Food, Environmental Degradation and Injustices*, Palgrave Studies in Green Criminology,
https://doi.org/10.1007/978-3-031-87900-5_1

Keywords Eco-justice · Environmental justice · Food systems · Sustainability · Equity

Introduction

Welcome! This book is going to look at a series of scenarios to explore the relationship between the foods which we as humans commonly consume, and the environmental degradation, and injustices which occur as a result. It takes as a starting point what White (2022) called "eco-justice" (2022, p. 19): a three-part system of justice that incorporates *environmental* justice, *ecological* justice, and *social* justice. Environmental justice being that which focuses primarily on the impact on humans, ecological justice being that which primarily focuses on the impact on habitat, and social justice being that which focuses either on individual non-human species or species as a whole. As such, it will be able to draw links between what White calls "illegal" harms (those which are contrary to the criminal law) and "legal" harms (those which are damaging and may be morally or ethically wrong but are not currently criminalised) (White 2022, p. 19). The book then explores these themes in relation to a range of different foodstuffs.

The idea for the book emerged over a series of walks in the Northamptonshire countryside with my wife Katie. We discussed the ethical complications of being vegetarian and wearing leather, and whether the best practicable environmental option (to borrow terminology from an old Royal Commission on Environmental Pollution Report—RECP 1988) was to wear plastic shoes and clothing made of recycled material. We then wondered what would happen if suddenly there were no more beef farms. Would cattle be bred for manure, meaning beef and leather would become by-products of the manure industry? The idea fascinated me, and over the next couple of years, the conceit for this book emerged.

The premise of the book is admittedly unusual, although similar approaches have been taken in fiction (for example, John Christopher's dystopian *Death of Grass* (1956) which traces the collapse of global society after a virus kills all forms of grass, including every grain consumed by humans). The book is not going to focus on strains of a particular foodstuff that are at risk of vanishing, such as Sarah Lohman does in relation to Carolina Gold rice or manoomin (Lohman 2023). It is instead an extended thought experiment, what Thompson called a story "cooked up to give us insights into an ethical problem" (2015, p. 4) predicated on the concept that the specific foodstuff or foodstuffs which are covered in each chapter immediately cease to exist in a tradeable form. The immediate and knock-on impacts of this will be traced as far as possible, including that in by-products and alternative food sources—if beef no longer exists, will people simply shift to eating sheep? The chapters will also consider any potential of some of the "novel" foods to offset some of the impacts of particular foodstuffs were to disappear. Novel foods are defined in the EU as "food that had not been consumed to a significant degree by humans in the EU before 15 May 1997. [...] 'Novel Food' can be newly developed, innovative food, food produced using new technologies and production processes, as well as food which is or has been traditionally eaten outside of the EU."[1] In addition to novel foods, chapters will consider what novel techniques exist, or are being developed, which might replace the "real" foodstuff. This will introduce us to ideas of laboratory-grown meat to replace beef, synthesised fish protein to replace salmon, and so on. In each chapter, having considered the justice issues and economics, each chapter will conclude whether, for example, the overall impact of a world without palm oil would be "better" or "worse" than one with palm oil.

Throughout the book, I have kept in mind that human beings have evolved over millions of years to be omnivores. Our physiology is designed to be able to cope with eating meat, fish, fruit, vegetables, grains, pulses, and pretty much anything we can grasp with our opposable thumbs. Any decision made by one person or a group of people not to eat particular foodstuffs is just that—a decision. The decision may have been triggered by medical circumstances (I am allergic to mussels, for example), religious beliefs (not eating pork or beef, not eating garlic or onions, for example), personal choice (flexitarianism, vegetarianism, or veganism for example), or simply by personal preference (not liking carrots for example). This book takes a neutral position on all of these—it is not a polemic against or in favour of particular decisions, it attempts to consider the evidence dispassionately and reach a series of balanced conclusions.

[1] Directorate General for Health and Food Safety (2023), Novel Food, European Commission, https://food.ec.europa.eu/safety/novel-food_en.

The book will introduce concepts with which you may not yet be entirely familiar, and so the next section will explain the key ideas that will be covered throughout the book.

Environmental Justice

Environmental justice is a phrase that is becoming quite widely used, but which has no precise definition. Some try and define it as a broad approach to policymaking, for example, the United States Environmental Protection Agency says it is "the just treatment and meaningful involvement of all people, regardless of income, race, color, national origin, tribal affiliation or disability" (EPA 2024). Others conceptualise environmental justice as a rights-based system that requires ecological unity and "ethical, balanced and responsible uses of land and renewable resources in the interest of a sustainable planet for humans and other living things" (Ramirez-Andreotta 2019). Still others see environmental justice as the *apophasis* of environmental *injustice*, which is to say, they define environmental justice as the absence of environmental injustice (Pellow 2007). The factor that unites the majority of the different definitions and approaches is that environmental policies, legislation, and decisions have had, are having, and will have disproportionate impacts on certain groups. Groups which are already advantaged in society are likely to see the greatest benefit, and groups which are already disadvantaged or marginalised are likely to see the greatest cost. These costs and benefits may manifest in financial, environmental, health, or other outcomes.

Environmental justice is not just occurring on a national scale. Internationally, the concept has been adapted to address global environmental challenges and inequalities. It is widely acknowledged that poorer nations often bear the brunt of the consequences of global environmental issues, such as climate change and biodiversity loss, which are disproportionately caused by wealthier nations. In the context of human food production, for example, areas of rainforest in the Brazilian Amazon are felled to produce pasture for cows, the meat from which is then exported. The poor farmers need the income to survive but suffer the consequences of deforestation and soil erosion (IPCC 2019). This means that if we are aiming to rectify historical injustices and promote equitable responsibilities, a global perspective on environmental justice is crucial in international negotiations and agreements (Parks and Roberts 2006).

Environmental justice is deeply rooted in human social and economic contexts, emphasising fairness in environmental policy and health impacts on human communities (Schlosberg 2004). With this specific focus on human communities and their possible right to a healthy environment, environmental justice addresses social inequalities that ecological and species justice (see below) may not tackle directly. For example, many environmental justice initiatives involve grassroots activism and legal battles aimed at preventing the siting of harmful industries in poor or minority areas (e.g., Bullard 1990), promoting community-led environmental stewardship, and influencing policy to ensure equitable access to natural resources.

While there is no uniformly accepted definition of environmental justice, the 1991 People of Color Environmental Leadership Summit created a set of seventeen "Principles of Environmental Justice":

(1) Environmental justice affirms the sacredness of Mother Earth, ecological unity and the interdependence of all species, and the right to be free from ecological destruction.
(2) Environmental justice demands that public policy be based on mutual respect and justice for all peoples, free from any form of discrimination or bias.
(3) Environmental justice mandates the right to ethical, balanced, and responsible uses of land and renewable resources in the interest of a sustainable planet for humans and other living things.
(4) Environmental justice calls for universal protection from nuclear testing, extraction, production, and disposal of toxic/hazardous wastes and poisons and nuclear testing that threaten the fundamental right to clean air, land, water, and food.
(5) Environmental justice affirms the fundamental right to political, economic, cultural, and environmental self-determination of all peoples.
(6) Environmental justice demands the cessation of the production of all toxins, hazardous wastes, and radioactive materials, and that all past and current producers be held strictly accountable to the people for detoxification and the containment at the point of production.
(7) Environmental justice demands the right to participate as equal partners at every level of decision-making, including needs assessment, planning, implementation, enforcement, and evaluation.

(8) Environmental justice affirms the right of all workers to a safe and healthy work environment without being forced to choose between an unsafe livelihood and unemployment. It also affirms the right of those who work at home to be free from environmental hazards.
(9) Environmental justice protects the right of victims of environmental injustice to receive full compensation and reparations for damages as well as quality health care.
(10) Environmental justice considers governmental acts of environmental injustice a violation of international law, the Universal Declaration On Human Rights, and the United Nations Convention on Genocide.
(11) Environmental justice must recognise a special legal and natural relationship of Native Peoples to the US government through treaties, agreements, compacts, and covenants affirming sovereignty and self-determination.
(12) Environmental justice affirms the need for urban and rural ecological policies to clean up and rebuild our cities and rural areas in balance with nature, honouring the cultural integrity of all our communities, and providing fair access for all to the full range of resources.
(13) Environmental justice calls for the strict enforcement of principles of informed consent, and a halt to the testing of experimental reproductive and medical procedures and vaccinations on people of colour.
(14) Environmental justice opposes the destructive operations of multinational corporations.
(15) Environmental justice opposes military occupation, repression and exploitation of lands, peoples and cultures, and other life forms.
(16) Environmental justice calls for the education of present and future generations which emphasises social and environmental issues, based on our experience and an appreciation of our diverse cultural perspectives.
(17) Environmental justice requires that we, as individuals, make personal and consumer choices to consume as little of Mother Earth's resources and to produce as little waste as possible, and make the conscious decision to challenge and reprioritise our

lifestyles to ensure the health of the natural world for present and future generations (PCELS 1991).

Although it does not appear as a specific principle in the list above, in the present work, considerations will also be given to the potential for exploitation of contractual, casual, and illegal workers at all stages of the food production process. A great deal of the food consumed in the UK is imported, and the UK Government estimates food imports were worth £48bn in 2020 (DEFRA 2023). While diversifying the location of food production has its advantages—it reduces the potential impact if one location is rendered temporarily underproductive due to climatic, political, or other conditions—it does mean that the workers are not all covered by the same legislative provisions about unionisation, health care, remuneration, or safety.

Each chapter in the book will consider the human health implications associated with each foodstuff. Is the food, like processed red meat, regarded as carcinogenic to humans by the International Agency for Research on Cancer (WHO 2023), for example, or does it contain useful fats such as the Omega-3 contained in salmon and other oily fish (NHS 2023). It will also look at whether there are differential health impacts on different socio-economic and other groups due to access to unprocessed, processed, or ultra-processed versions.

Ecological Justice

Much like environmental justice, ecological justice is defined in several different ways. White (2022) broadly categorises ecological justice as a system in which the victim of injustice is a specific ecosystem (2022, p. 18). Baxter (2004) suggests that "the concept of ecological justice imposes obligations on human moral persons to act on behalf of the claims in justice of non-person organisms to their fair share of environmental resources" (2004, p. 80), which requires an ab initio recognition that non-human species are entitled to a "fair share" of resources. Bhakuni (2021) takes this idea further, by suggesting that one way of fostering ecological justice would be to "treat the health of biotic and abiotic environments non-instrumentally" (2021, p. 7)—in other words the health of living (including non-human) and non-living environments regardless of the extent to which this health impacts humans. The approach taken in the current text will be to consider ecological justice as that which protects the inherent rights of the ecosystems impacted by food production, which can have profound and often detrimental impacts on specific ecosystems. The focus is on inherent rights, rather than those granted by human actors, although it recognises that, as Bookchin (1993) put it three decades ago, "all ethical systems… are formulated by human beings in distinctly cultural situations" (1993, pp. 47–48).

At a pragmatic level, each chapter will therefore consider the ecological issues that arise as a result of the production (etc.) of the specified foodstuff. This will involve different specific issues in different chapters, but will broadly cover *space*, *place*, and *waste*.

Space. Whatever the foodstuff, it requires physical space to be grown. If the space is on land, as with meat production, then there may be issues around deforestation, hedgerow destruction, monoculture, and human-non-human conflict. One stark illustration is the expansion of agricultural land in the Amazon rainforest. This activity has led to extensive deforestation, which not only destroys habitats but also contributes to climate change due to both the release of stored carbon dioxide and the reduction in carbon sequestration (Nepstad et al. 2008). The soybean farms and cattle ranches that replace the forest are less biodiverse and offer fewer ecological services than the original rainforest (Fearnside 2005). Animals and crops both need water to thrive, and increased abstraction has negative impacts on aquifers and water tables, as can be seen currently in Spain and the United States, or the catastrophic reduction in the size of the Aral Sea in Kazakhstan and Uzbekistan (Miklin et al. 2014), for example. Aquaculture too, particularly shrimp farming, has substantial impacts on coastal ecosystems (Barbier 2016). If the space is on, in, or under water, then there may be issues of infringement on protected spaces, the removal of bodies of water from other uses, and, again, an impact on those who previously occupied the space. In Southeast Asia, the growth of shrimp farming has led to the destruction of mangrove forests, which are crucial for carbon sequestration, shoreline stabilisation, and as nurseries for fish. The loss of mangroves to aquaculture not only impacts biodiversity but also leaves coastal areas more vulnerable to storms and flooding (Barbier 2016). The differential impact of these water shortages on different people will also link back to issues of environmental justice.

Place. The physical space that all foodstuffs need does not exist in the abstract. It is located in a specific country. The impact of transportation of the foodstuff from "field to fork" is one of the factors that causes discussion among those looking at sustainability. From a strictly ecological perspective, the use of commercial flights to move fresh produce from the zone of production to the zone of consumption can be seen as a major and unnecessary contributor to the anthropogenic emissions that are causing the climate crisis (Sanderon et al. 2019). From a social sustainability perspective, the same trade provides a valuable source of income for marginalised peoples. However, Indigenous rights to land and resources are often compromised by large-scale agricultural projects, which can lead to conflicts and displacement (Schlosberg 2007). Some try to balance these competing interests by trying the disingenuous argument that since the flight was going anyway, adding fresh produce into the hold has a marginal impact on climate at best, and adds to social sustainability, and so is a net sustainable trade (Laryea et al. 2023).

Waste. Rainwater which runs off fields used for animal or crop production is likely to contain higher levels of nitrogen than normal, especially if the crops have been treated with nitrogen-based fertilisers (Even and Deane 2018) or the animals are present in sufficient numbers that their bodily waste exceeds the capacity of the land to absorb it. Aquaculture too, whether land based or offshore, also causes nitrogen-rich waste due to the density of the fish contained within the tanks or cages (Troell et al. 2014). If river water or sea water is too rich in nitrogen, this can cause a phenomenon known as Eutrophication. This can cause toxic algal blooms and the death of marine life, particularly in estuaries and coastal areas (NOAA 2024). The death of fish affects the livelihoods of close to half a billion people globally who depend upon small-scale fisheries for their livelihoods (UN 2023), and this links back to the ideas of environmental justice explored above.

Species Justice

The third part of White's division of "eco-justice" is species justice. Ethically, the discussions for this part of each chapter will focus on the welfare and sentience of the non-human animals which *are* the foodstuff, in addition to the subsequent effects of producing the specified foodstuff on other species. The question of whether it would be "better" for a non-human animal that it is born, lives, dies, and is eaten, or never to have been born at all has fascinated philosophers for centuries. UCL Professor of Philosophy Nick Zangwill argues forcefully that humans "owe it to animals to eat them" because by doing so we are continuing a "long-standing cultural institution that is a mutually beneficial relationship" (Zangwill 2022). He does introduce two strong caveats to his argument, which are that the animals in question have lived "good lives" free from pain and suffering, and that "Of course, the practice does not benefit a particular animal that we eat at the time that we eat it" (Zangwill 2021). Lomasky too argues that although animals "are exploited by being made agricultural commodities, but that exploitation is not necessarily bad for them. Because we prize them, they do better than if we did not, although what we prize them for is primarily their savor and not their well-being" (Lomasky 2013, p. 200).

Whatever the merits of this argument Zangwill acknowledges that it is based on an idealistic premise that farmed animals can, or do, live "good lives." Lomasky (though not in specific terms) suggests the current meat production system is environmentally unjust, and that if the system were reformed so that "those who reap the benefits also bear the costs" (2013, p. 198) that would bring benefit to both human and non-human animals. Porcher (2011) looked at the lived experiences of animals in "concentrated animal feeding operations" (otherwise known as intensive farming where the animals spend 45 days or more indoors) (ERS 2024), and concluded that this has profound impacts and devalues the life and dignity of the animals (Porcher 2011).

In practice, animal welfare standards differ from region to region, and may be legally enforceable or simply coded as "best practice." This divergence is an example of what White has called legal and illegal wrongs (White 2022).

In England and Wales, for example, section 9 of the Animal Welfare Act 2006 appears at first glance to be a strong example of legally imposed obligations. The section states that "A person commits an offence if he does not take such steps as are reasonable in all the circumstances to ensure that the needs of an animal for which he is responsible are met to the extent required by good practice." On the plus side, this does impose a legal duty to ensure the welfare of animals, but this is weakened by what I see as the three main *lacunae* in the section:

- The needs of the animal are only required to be met "to the extent required by good practice" (s9(1)). Good practice is not defined by statute, and although the power issue and revise codes of practice (which are analogous to "good practice") are delegated to "the

Appropriate National Authority" (in England this is the Department of the Environment, Food and Rural Affairs and in Wales it is the Welsh Assembly), there is merely a discretionary legal power to do this, rather than a mandatory legal duty. Up to 2024, Defra had issued Codes of Practice about the keeping of dogs, cats, horses, and privately kept primates (all in 2010). The Welsh Assembly, by contrast, had issued them for Cattle (2010), Sheep (2010), Rabbits (2011), Game Birds reared for sport (2011), Pigs (2014), Cats (2008/2019), Horses (2018), Dogs (2008/2019), Animal Welfare Establishments (2020), Laying Hens (2020), and Meat Chicken (2020). Despite the species themselves being identical in the two countries, the legal standards for their treatment in England and Wales therefore may be different. This difference will be magnified when looking at wider jurisdictions.
- The Act only applies to vertebrates other than man (s1(1)), but does not apply to "anything which occurs in the normal course of fishing" (s59). There have been discussions about adding some decapod crustaceans and cephalopod molluscs to the list of protected creatures based on their sentience (Birch et al. 2021) but, to date, this has not happened in the UK. The exclusion of fishing (recreational and commercial) from the Act was deliberate and was the subject of some specific debate during the passage of the Act. However, many fishing techniques, such as trawling and the use of gillnets, also cause extensive bycatch—unintended capture of non-target species, including dolphins, sea turtles, and seabirds, leading to injuries or death (Alverson et al. 1994).
- Section 4 offences (causing unnecessary suffering) only apply to "protected animals" defined in s2 as commonly domesticated in the British Isles, under the control of man—temporarily or permanently—and not living wild. So, a wild deer, injured but not killed by a crossbow bolt, would not be covered by the legislation, despite the obvious levels of "unnecessary suffering" caused.

The chapter acknowledges that, as Mariam Stamp Dawkins pointed out (2016), there is an inviolable link between the welfare of farm animals and the cost to farmers, and thus consumers. Farming and production do not exist in a vacuum, and the book will investigate the consequences of price rises for food amidst a cost-of-living crisis.

Economics

Each chapter will end with a consideration of the economic impacts of the sudden cessation of particular food products. Food production significantly influences local, regional, and global economies, affecting everything from employment and income levels to trade balances and economic development. The multifaceted impacts of food production can be both positive and negative, shaping economies through the dynamics of agricultural productivity, trade policies, market fluctuations, and socio-economic development. Understanding these impacts, which can be felt at local, regional, and global levels, requires a comprehensive look at how food production interacts with economic structures and policies.

At the local level, agriculture is often a cornerstone of the economy, particularly in rural areas in the Global South, where it can be the primary source of income and employment. This has strong links to UN Sustainable Development Goal 2 (End Hunger, achieve food security and improve nutrition and promote sustainable agriculture), and particularly Target 2.3, which calls for agricultural productivity and incomes of small-scale producers (in particular women, Indigenous peoples, family farmers, pastoralists, and fishers) to double by 2030 (UN DESA 2024). On the positive side, the presence of robust food production can stimulate local economies through direct employment in farming, as well as indirectly in associated industries such as equipment manufacturing, food processing, and retail. A well-developed agricultural sector can improve infrastructure, better services, and increase economic activity in rural areas. However, the benefits can be unevenly distributed, and intensive farming practices may have negative long-term effects. For example, over-reliance on a single crop can make local economies vulnerable to price shocks and crop failures due to disease or adverse weather conditions. Moreover, the intensification of agriculture can lead to environmental degradation, such as soil depletion and water scarcity, which in turn can undermine future agricultural productivity and economic stability (Tilman et al. 2002).

Regionally, food production shapes economic integration and development through trade networks and agricultural policies. Regions abundant in agricultural resources, or with conditions better suited to growing specific foodstuffs, often export food to less fertile areas, influencing regional trade balances and economic relationships. For instance, the Midwest region of the United States is a significant exporter of corn and soybeans, which supports regional and national economic strength through export revenues.

Agricultural policies, such as subsidies and support for specific crops, also have profound effects on regional economies. These policies can dictate which crops are grown and which practices are adopted, potentially skewing regional development towards certain agricultural sectors while neglecting others. Additionally, investments in agricultural research and development can enhance regional competitiveness by improving crop yields and reducing vulnerability to pests and diseases (Alston et al. 2000).

The UN Food and Agricultural Organization put the importance of the global food trade succinctly, saying:

> Trade in food and agriculture has been an essential part of our history and is important to societies. Countries engage in trade to export what they can produce at a lower cost relative to other countries while importing what is relatively more expensive to produce domestically. For a country, many factors can influence trade in food and agricultural products, but the most influential factor is comparative advantage – a country's ability to produce a particular good at a lower opportunity cost than its trading partners. (UN FAO 2023, p. xii)

The global food market affects international trade, economic stability, and nations' development trajectories. Countries that are major food exporters, such as Brazil and the United States, play pivotal roles in global food security and economic stability. The relative fragility of the global food market was revealed in 2022 when in the immediate aftermath of the Russian invasion of Ukraine, a country that exported US$27.8bn of wheat, corn, and other agricultural products in 2021 (FAS 2022), global wheat futures prices rose by around 40% (Devadoss and Ridley 2024). Other factors such as climate change or trade policies can have ripple effects across the world, influencing global commodity prices and economic policies (FAO 2023).

Moreover, food production is closely linked to other global issues of environmental injustice, such as poverty and development. In developing countries, improving agricultural productivity is often a key strategy for economic development, as it can reduce poverty directly by increasing farm incomes and indirectly by decreasing food prices, which benefits all consumers. The World Bank has emphasised the role of agriculture in economic development and poverty reduction in low-income countries (World Bank 2007).

There is a network of business operations and regulatory frameworks including agricultural businesses, pharmaceutical companies, feed suppliers, and governmental bodies that support the mass production and consumption of animal products. Noske (1989) first called this network the "animal-industrial complex," mirroring the idea of the "military-industrial complex" which has been around since the early 1960s (Eisenhower 1961). Twine (2012), Nibert (2013), Maher et al. (2017), Beirne and Lynch (2023), and others argue that this system, and the intensive farming practices which it finds economically most suitable, perpetuates animal suffering on a massive scale, contributes significantly to environmental degradation, is speciesist in nature, and often overlooks public health in favour of economic gains. Stallwood (2014) further suggests that the animal-industrial complex, far from being deterred by protests, is actually emboldened by the focus on personal choice promulgated by the animal rights movement (2014, p. 302).

The shift towards large-scale industrial agriculture has raised concerns about the sustainability of food production systems. This model of production often involves significant environmental costs, including greenhouse gas emissions and biodiversity loss, which could pose long-term risks to economic stability. The Woodland Trust estimates that "11,800 miles of have disappeared [in Britain] since 1950, due largely to intensification of agriculture" (Woodland Trust 2024) and with it the biodiversity that it contained. Rockström et al. (2017) argue that the transition to more sustainable agricultural practices is seen as essential for long-term economic health, in addition to the environmental and social benefits.

The impact of food production and the animal-industrial complex on local, regional, and global economies is profound and complex. While it is a source of income and livelihood for millions of people worldwide, it also poses significant challenges and risks to ethics, environmental sustainability, and public health that need to be managed through thoughtful economic and environmental policies. Ensuring the sustainability of food production systems is crucial for future economic stability and development at all levels of the economy.

Conclusion

This book sets out to examine the relationship between food production and justice—environmental, ecological, and species based—through the lens of thought experiments where specific food commodities cease to exist in tradeable form. It is intended to provoke reflection and discussion on the often-unseen ramifications of our food systems, using justice frameworks to navigate complex global interdependencies. The remainder of this book is broken into four chapters, each of which focuses on two linked foodstuffs. They foodstuffs were chosen because they are either the most commonly consumed example of a category, or they have the largest environmental impact, or both.

Chapter 2 (Meat). This chapter investigates the environmental, social, and economic consequences of porcine and bovine meat production. It begins by exploring the significance of these meats in global diets, considering their nutritional benefits alongside health risks such as links to chronic diseases. It critiques industrial farming practices, especially concentrated animal feeding operations (CAFOs), for their disproportionate environmental and health impacts on marginalised communities. The chapter also discusses the ecological consequences of land conversion for grazing, including deforestation and biodiversity loss. By addressing economic inequalities and animal welfare concerns, it concludes with an evaluation of whether alternative protein sources could mitigate these harms while sustaining food security.

Chapter 3 (Fruit and Vegetables). Focusing on Tomatoes and Potatoes, this chapter examines their centrality to food systems and cultural heritage. It unpacks the environmental costs of intensive tomato and potato cultivation, including overuse of water and over-reliance on pesticide, while also exploring monoculture farming's threat to biodiversity. The analysis extends to social justice, highlighting the exploitation of migrant labour in key producing regions. The chapter emphasises the economic ripple effects of losing these crops from global trade, particularly for smallholder farmers and low-income nations. It proposes that a shift towards diversified local food systems could support more equitable and sustainable agricultural practices.

Chapter 4 (Fish and Seafood). This chapter examines aquaculture's dual role as a solution to overfishing and a contributor to environmental degradation. It focuses on farmed prawns, salmon, and tilapia, exploring their environmental impacts, including water pollution, habitat destruction, and the spread of antibiotic resistance. Social justice issues, such as labour exploitation and health risks for workers, are scrutinised. The chapter also assesses ecological justice concerns, including the disruption of aquatic ecosystems and the carbon footprint of aquaculture practices. It discusses whether ceasing aquaculture production could reconcile these injustices while ensuring sustainable seafood supplies.

Chapter 5 (Palm Oil and Soya). The final substantive chapter shifts attention to two commodity crops that epitomise the tensions between industrial agriculture, global trade, and sustainability. Palm oil and soya are critical to numerous industries but exact significant environmental and social costs. The chapter examines the deforestation and biodiversity loss driven by their cultivation, as well as labour exploitation and the marginalisation of Indigenous communities. Sustainability initiatives like the Roundtable on Sustainable Palm Oil are evaluated for their effectiveness in mitigating these harms. The chapter concludes by exploring alternative crops and technologies that might reduce reliance on palm oil and soya while addressing systemic inequities in global supply chains.

The themes explored in this book underscore an urgent imperative: to rethink and reshape food systems in ways that prioritise equity, sustainability, and resilience. The food commodities we rely on are entangled in vast networks of social, economic, and ecological dynamics, revealing the profound connections between personal choices and global consequences. By imagining the absence of specific foodstuffs, this book hopes to highlight the vulnerabilities and inequities embedded in current agricultural practices. The justice frameworks provide tools for critically evaluating these practices and imagining alternatives that respect both human life and non-human life. Whether through reducing meat consumption, adopting sustainable aquaculture, or supporting local crop diversity, there are tangible pathways to fostering a more just and sustainable food system.

It does not contain all of the answers—indeed, it does not even ask all of the questions, but hopefully you will find that the following pages raise questions about the industrial food production system, and possibly about your own practices.

Sources

Alston, J.M., C. Chan-Kang, M.C. Marra, P.G. Pardey, P.G., and T.J. Wyatt. 2000. A meta-analysis of rates of return to agricultural R&D: ex pede

Herculem?, Research reports 113, International Food Policy Research Institute (IFPRI).

Alverson, D.L., M.H. Freeberg, S.A. Murawski, and J.G. Pope. 1994. A global assessment of fisheries bycatch and discards, FAO Fisheries Technical Paper 339, United Nations Food and Agriculture Organisation. https://www.fao.org/4/t4890e/T4890E00.htm

Barbier, E.B. 2016. The protective service of mangrove ecosystems: A review of valuation methods. *Marine Pollution Bulletin* 109 (2): 676–681. https://doi.org/10.1016/j.marpolbul.2016.01.033.

Baxter, B. 2004. *A theory of ecological justice*. London: Taylor & Francis Group.

Beirne, P., and M.J. Lynch. 2023. On the geometry of speciesist policing: The federal bureau of investigation's animal cruelty data. *International Journal for Crime, Justice and Social Democracy* 12 (2): 137–149. https://doi.org/10.5204/ijcjsd.2631.

Bhakuni, H. 2021. Beyond anthropocentrism: Health rights and ecological justice. *Health and Human Rights Journal* 23 (2): p7-12.

Birch, J., C. Burn, A. Schnell, H. Browning, and A. Crump. 2021. Review of the evidence of sentience in cephalopod molluscs and decapod crustaceans. LSE Consulting, London School of Economics. https://www.lse.ac.uk/business/consulting/assets/documents/Sentience-in-Cephalopod-Molluscs-and-Decapod-Crustaceans-Final-Report-November-2021.pdf.

Bookchin, M. 1993. Deep ecology, anarcho-syndicalism, and the future of anarchist thought. In *Deep ecology & anarchism: A polemic*. London: Freedom Press.

Bullard, R. 1990. *Dumping in Dixie: Race, class, and environmental quality*. Westview Press.

Christopher, J. 1956. *The death of grass*. London: Michael Joseph.

Dawkins, M.S. 2016. *Why animals matter: Animal consciousness, animal welfare, and human well-being*. Oxford: OUP.

DEFRA, 2023. United Kingdom Food Security Report 2021: Theme 2: UK food supply sources, updated 2023. Department for Environment Food & Rural Affairs. https://www.gov.uk/government/statistics/united-kingdom-food-security-report-2021/.

Devadoss, S., and W. Ridley. 2024. Impacts of the Russian invasion of Ukraine on the global wheat market. *World Development* 173: 106396. https://doi.org/10.1016/j.worlddev.2023.106396.

Directorate General for Health and Food Safety. 2023. Novel food. European Commission. https://food.ec.europa.eu/safety/novel-food_en.

Eisenhower, D.D. 1961. Milestone documents: President Dwight D. Eisenhower's farewell address, US National Archives. https://www.archives.gov/milestone-documents/president-dwight-d-eisenhowers-farewell-address.

EPA. 2024. Learn about environmental justice. United States Environmental Protection Agency. https://www.epa.gov/environmentaljustice/learn-about-environmental-justice.

ERS. 2024. Agricultural trade. US Department of Agriculture, Economic Research Service. https://www.ers.usda.gov/data-products/ag-and-food-statistics-charting-the-essentials/agricultural-trade/.

Even, H., and F. Deane. 2018. The control of nutrient run-off from agricultural areas: Insights into governance from Australia's sugarcane industry and the Great Barrier Reef. *Transnational Environmental Law* 7 (3): 451–468. https://doi.org/10.1017/S2047102518000018.

FAS. 2022. Ukraine agricultural production and trade. Foreign Agricultural Service, US Department of Agriculture. https://fas.usda.gov/sites/default/files/2022-04/Ukraine-Factsheet-April2022.pdf.

Fearnside, P.M. 2005. Deforestation in Brazilian Amazonia: History, rates, and consequences. *Conservation Biology* 19 (3). https://doi.org/10.1111/j.1523-1739.2005.00697.x.

IPCC. 2019. Climate change and land an IPCC special report on climate change, desertification, land degradation, sustainable land management, food security, and greenhouse gas fluxes in terrestrial ecosystems. InterGovernmental Panel on Climate Change. https://www.ipcc.ch/site/assets/uploads/2019/11/SRCCL-Full-Report-Compiled-191128.pdf.

Laryea, E., A. Hosseinian-Far, and S. Derrick. 2023. Climate justice implications of banning air-freighted fresh produce. *Logistics* 7 (4): 78. https://doi.org/10.3390/logistics7040078.

Lohman, S. 2023. *Endangered eating: America's vanishing foods*. London: W. W. Norton & Co Ltd.

Lomasky, L. 2013. Is it wrong to eat animals? *Social Philosophy & Policy* 30 (1–2): 177–200. https://doi.org/10.1017/S0265052513000083.

Maher, J., H. Pierpoint, and P. Beirne. 2017. *The Palgrave international handbook of animal abuse studies*. London: Palgrave MacMillan.

Miklin, P., N.V. Aladin, and I. Plotnikov, eds. 2014. *The Aral Sea: The devastation and partial rehabilitation of a Great Lake*. Heidelberg: Springer, Berlin.

Nepstad, D.C., C.M. Stickler, B. Soares-Filho, and F. Merry. 2008. Interactions among Amazon land use, forests, and climate: Prospects for a near-term forest tipping point. *Philosophical Transactions of the Royal Society B* 363 (1498): 1737–1746. https://doi.org/10.1098/rstb.2007.0036.

NHS. 2023. Fish and shellfish. National Health Service. https://www.nhs.uk/live-well/eat-well/food-types/fish-and-shellfish-nutrition/.

Nibert, D. 2013. *Animal oppression and human violence: Domesecration, capitalism, and global cconflict*. New York: Columbia University Press.

NOAA. 2024. What is eutrophication? National Ocean Service; National Oceanic and Atmospheric Administration. https://oceanservice.noaa.gov/facts/eutrophication.html.

Noske, B. 1989. *Humans and other animals: Beyond the boundaries of anthropology*. London: Pluto Press.

Parks, B.C., and J.T. Roberts. 2006. Environmental and ecological justice. In *Palgrave advances in international environmental politics*, ed. by M.M. Betsill, K. Hochstetler, and S. Demitris. Basingstoke: Palgrave Macmillan.

PCELS. 1991. The principles of environmental justice (EJ). People of Color Environmental Leadership Summit, Washington DC, EJnet.org. https://www.ejnet.org/ej/principles.pdf.

Pellow, D.N. 2007. *Resisting global toxics: Transnational movements for environmental justice*. Cambridge, MA: MIT Press.

Porcher, J. 2011. The relationship between workers and animals in the pork industry: A shared suffering. *Journal of Agricultural and Environmental Ethics* 24 (1): 3–17. https://doi.org/10.1007/s10806-010-9232-z.

Ramirez-Andreotta, M. 2019. Environmental justice. In Environmental and pollution science, 3rd ed, ed. by M.L. Brusseau, I.L. Pepper, and C.P. Gerba. London: Elsevier.

RCEP. 1988. *Best practicable environmental option: 12th Report of the Royal Commission on Environmental Pollution*. London: HMSO.

Rockström, J., J. Williams, G. Daily, A. Noble, N. Matthews, L. Gordon, H. Wetterstrand, F. DeClerck, M. Shah, P. Steduto, C. de Fraiture, N. Hatubi, O. Unver, J. Bird, L. Sibanda, and J. Smith. 2017. Sustainable intensification of agriculture for human prosperity and global sustainability. *Ambio* 46 (1): 4–17. https://doi.org/10.1007/s13280-016-0793-6.

Sanderon, V., N. Bamber, and N. Pelletier. 2019. Cradle-to-market life cycle assessment of Okanagan (Canada) cherries: Helicopters, seasonal migrant labour and flying fruit. *Journal of Cleaner Production* 229: 1283–1293. https://doi.org/10.1016/j.jclepro.2019.04.398.

Schlosberg, D. 2004. Reconceiving environmental justice: Global movements and political theories. *Environmental Politics* 13 (3): 517–540. https://doi.org/10.1080/0964401042000229025.

Schlosberg, D. 2007. *Defining environmental justice: Theories, movements, and nature*. Oxford: OUP.

Stallwood, K. 2014. Animal rights: Moral crusade or social movement? In *Critical animal studies: Thinking the unthinkable*, ed. by J. Sorensen. Toronto: Canadian Scholars.

Thompson, P.B. 2015. *From field to fork: Food ethics for everyone*. Oxford: OUP.

Tilman, D., K.G. Cassman, P.A. Matson, R. Naylor, and S. Polasky. 2002. Agricultural sustainability and intensive production practices. *Nature* 418 (6898): 671–677. https://doi.org/10.1038/nature01014.

Troell, M., R.L. Naylor, M. Metian, and A. de Zeeuw. 2014. Does aquaculture add resilience to the global food system? *Proceedings of the National Academy of Sciences* 11 (37): 13257–13263. https://doi.org/10.1073/pnas.1404067111.

Twine, R. 2012. Revealing the 'animal-industrial complex'—A concept & method for critical animal studies? *Journal for Critical Animal Studies* 10 (1): 12–39.

UN. 2023. Goal 14: Conserve and sustainably use the oceans, seas, and maritime resources. United Nations Sustainable development Goals. https://www.un.org.sustainabledevelopment/oceans.

UN DESA. 2024. Sustainable development goal 2. United Nations Department of Economic and Social Affairs. https://sdgs.un.org/goals/goal2.

UN FAO. 2023. The state of agricultural commodity markets 2022. The geography of food and agricultural trade: Policy approaches for sustainable development. Food and Agriculture Organization of the United Nations. https://openknowledge.fao.org/handle/20.500.14283/cc0471en.

White, R. 2022. *Theorising green criminology*. Abingdon: Routledge.

WHO. 2023. Red and processed meat in the context of health and the environment: Many shades of red and green. World Health Organisation. https://www.who.int/publications/i/item/9789240074828.

Woodland Trust. 2024. Hedgerows. The Woodland Trust. https://www.woodlandtrust.org.uk/trees-woods-and-wildlife/habitats/hedgerows/.

World Bank. 2007. *World development report 2008: Agriculture for development*. Washington, DC: World Bank. http://hdl.handle.net/10986/5990. License: CC BY 3.0 IGO.

Zangwill, N. 2021. Our moral duty to eat meat. *Journal of the American Philosophical Association* 7 (3): 295–311. https://doi.org/10.1017/apa.2020.21.

Zangwill, N. 2022. Why you should eat meat. Aeon. https://aeon.co/essays/if-you-care-about-animals-it-is-your-moral-duty-to-eat-them.

CHAPTER 2

Meat

Abstract This chapter examines the global trade and production of pig and cow meat, highlighting their significant dietary roles alongside the environmental, social, and health challenges they pose. Pig meat and cow meat are staples of global diets, providing essential protein and micronutrients, but their production is linked to extensive environmental degradation, including deforestation, biodiversity loss, and greenhouse gas emissions. The chapter discusses concentrated animal feeding operations (CAFOs), emphasising their contribution to ecological harm and the disproportionate health burdens on nearby marginalised communities. Health-related concerns include the association of red and processed meat with diseases such as cardiovascular disorders, cancer, and metabolic syndrome, contrasting with their nutritional benefits. Social equity issues are explored, particularly land dispossession and the economic inequalities perpetuated by industrial meat systems. The chapter also analyses economic dynamics, underscoring how profits often bypass affected communities while benefiting multinational corporations. Concluding with potential alternatives such as lab-grown meat and plant-based innovations, the chapter calls for a transition to equitable and sustainable food systems that balance nutrition, justice, and environmental preservation.

Keywords Industrial farming · Environmental degradation · Biodiversity loss · Red meat consumption · Sustainable food systems

© The Author(s), under exclusive license to Springer Nature Switzerland AG 2025

S. Sneddon, *Food, Environmental Degradation and Injustices*, Palgrave Studies in Green Criminology,
https://doi.org/10.1007/978-3-031-87900-5_2

Introduction

This chapter focuses on aspects of the global trade in meat. The global trade and consumption of meat remain pivotal to discussions about diet, environment, and social equity, reflecting a complex interplay of culture, health, and industry. This chapter examines the production and consumption of pig and cattle meat, delving into their significance within dietary patterns and their broader implications. Pig meat, despite being prohibited in certain cultural and religious contexts, is the second most consumed meat worldwide, accounting for more than a third of all meat consumption at approximately 120 million tons in 2023 (FAO 2024). Cow meat, which is consumed at a rate of 80 million tons a year (FAO 2024), represents a leading contributor to greenhouse gas emissions and environmental degradation. The chapter employs the United Nations Food and Agriculture Organization's terminological framework, using "pig" and "cow" to address these products collectively, irrespective of their varied culinary forms like pork, ham, beef, or veal. Pig meat and cow meat[1] are classified by the World Health Organisation as "red meat" because they contain higher levels of myoglobin than poultry, fish, and seafood (WHO 2023).

The discussion begins by situating pig and cow meats within global consumption trends, highlighting their dietary prominence and nutritional attributes. Both are rich sources of protein and essential micronutrients, including B vitamins and heme iron, yet their health impacts are far from unequivocal. Processed meats, in particular, are classified as carcinogenic, and high consumption levels of red meat have been associated with a range of health risks, from cardiovascular diseases to metabolic disorders. These trends raise significant questions about the role of these meats in modern diets, especially in regions where consumption patterns far exceed nutritional necessity.

However, the implications of pig and cow meat consumption extend well beyond individual health. Their production systems, particularly the rise of concentrated animal feeding operations (CAFOs), are linked to environmental injustice, biodiversity loss, and social inequality. The chapter explores how industrial-scale farming disrupts ecosystems, exacerbates climate change, and imposes severe health burdens on nearby

[1] The WHO categorisation (which uses porcine (pig) and bovine (cow)) also includes ovine (sheep), caprine (goat), equine (horse), and cervine (deer) meats (WHO 2023).

communities. CAFOs disproportionately affect marginalised populations, often situated in low-income or minority-dense regions, and contribute to zoonotic disease outbreaks due to overcrowded and unsanitary conditions.

Environmental degradation is another critical dimension of the discussion. Cow farming, in particular, drives deforestation, especially in biodiverse regions like the Amazon rainforest, where land clearance for grazing directly threatens thousands of species and accelerates carbon emissions. Similarly, the vast land, water, and resource demands of livestock farming underscore its unsustainable ecological footprint. The consequences of these practices resonate globally, from altered hydrological cycles to increased vulnerability to extreme weather events, creating a feedback loop of environmental harm.

The chapter further interrogates the economic dynamics underpinning the global trade in pig and cow meats. This industry, valued at hundreds of billions of dollars annually, supports intricate supply chains spanning production, processing, transportation, and retail. Yet, it also entrenches inequities, as the wealth generated often bypasses the rural and vulnerable communities most affected by industrial farming practices. Moreover, the infiltration of Organised Criminal Groups (OCGs) into certain facets of livestock farming exemplifies the economic complexities and ethical dilemmas tied to this sector.

By locating these issues within the broader framework of justice—environmental, ecological, and species based—the chapter evaluates the welfare of farmed animals, the biodiversity impacts of land-use change, and the systemic inequalities perpetuated by meat production and consumption. While acknowledging the nutritional and economic contributions of pig and cow meat, the chapter calls attention to the pressing need for a more equitable and sustainable approach to food systems.

Through these lenses, the chapter not only critiques the status quo but also sets the stage for exploring alternatives to pig and cow meat consumption, from dietary shifts and plant-based innovations to lab-grown meat. The chapter concludes with a balance assessment of the negative and positive impacts of an immediate cessation of the trade in cow and pig meat.

Environmental Justice

As we have seen, both pig meat and cow meat are significant components of diets worldwide. However, the production and consumption of these meats have been linked to various issues relating to environmental injustice, particularly in relation to human health issues and social inequalities.

Processed meat is identified by the International Agency for Research on Cancer (IARC), a part of the World Health Organization, as a Group 1 threat to human health (it is *definitely* carcinogenic) and unprocessed red meat is listed as a Group 2A threat (it is *probably* carcinogenic). In 2021, Papier et al. linked the consumption of red meat (all red meat, not just pig and cow meat) to an increased risk of heart disease, pneumonia, diverticulitis, and diabetes (Papier et al. 2021). According to a report by Bouvard et al. (2015), substances used in the processing of meat such as nitrates or nitrites and the presence of chemicals formed during meat processing and cooking, such as polycyclic aromatic hydrocarbons (PAH) and heterocyclic aromatic amines (HAA), are of particular concern, as they are known carcinogens.

A review by Wang and Beydoun (2009) indicates a positive correlation between high meat consumption and obesity rates in different populations in the United States, highlighting the potential benefits of dietary adjustments in meat consumption. One of the major health concerns associated with the consumption of cow meat is its contribution to cardiovascular disease (CVD) (Hu et al. 2024). Cow meat, particularly processed forms, is high in saturated fats and cholesterol, which are known risk factors for CVD (Clifton and Keogh 2017). A meta-analysis by Micha et al. (2010) revealed a significant association between the consumption of processed red meat and an increased risk of coronary heart disease (CHD) and type 2 diabetes, underscoring the need for dietary moderation in the intake of such foods. Li et al. (2024) also noted an increase in diabetes with higher consumption of processed and unprocessed red meat and poultry. Eating pig meat, especially when cured, has also been linked to an increased risk of various types of cancer, particularly colorectal cancer (Fiego et al. 2018). Equally, various studies have linked the intake of bioavailable iron (sometimes called "heme iron") from all red meat with increased rates of cancer, metabolic syndrome, and CVD (De Olivera Otto et al. 2012; Sasso and Latella 2018). Bovine spongiform encephalopathy (BSE), also known as mad cow disease, is a zoonotic neurological disorder in cattle that can affect humans, causing variant Creutzfeldt-Jakob disease, a fatal brain disorder (Collinge 2001).

However, there are some more positive health impacts of eating pig and cow meats. Both are excellent sources of high-quality protein, which is essential for muscle growth, repair, sustaining muscle mass in ageing populations, and overall bodily functions (Kreider and Campbell 2009; Wolfe 2017). Proteins from these meats contain all nine essential amino acids needed by the human body, making them complete proteins. In addition to protein, both are rich in various micronutrients. They are important sources of B vitamins, particularly vitamin B12, which is crucial for red blood cell formation, neurological function, and DNA synthesis and is not found naturally in most plant foods (Zeuschner et al. 2013). Meats also provide vitamin B3 (niacin), which plays a role in the metabolism of sugars and fatty acids, but which is also produced by plants including peas, mushrooms, and tomatoes (Wagner et al. 1947). Pig and cow meats also contain vitamin B6, which is important for brain development and function (Bekdash 2023). Again, vitamin B6 is found in starchy vegetables such as potatoes, sweet potatoes, and bananas (NIH 2024).

In addition to B vitamins, red meat is one of the best dietary sources of heme iron. Iron, despite the risks posed by over-consumption identified above, is vital for transporting oxygen in the blood and for energy production. Zinc, another mineral abundantly found in these meats, supports immune function, wound healing, and DNA synthesis (Hunt 2003). Meat is the primary dietary source of creatine, a compound important for energy production in muscle cells (Wu 2020) and carnosine, acts as an antioxidant, and has a role in muscle health (Baguet et al. 2010). A vegetarian diet appears not to affect carnosine (Blancquaert et al. 2018), and consumption of tofu, lentils, and nuts can help the human body produce creatine, but not sufficiently to replace that gained from meat consumption (Blancquaert et al. 2018).

However, it is not just the consumers of pig and cow meat who may suffer health impacts. The health implications for those living near concentrated animal feeding operations (CAFOs) are also a significant concern. Mirabelli et al (2006) and Schultz et al. (2019) found indicated increased incidence of respiratory problems such as asthma and nasal allergies among those who lived closer to CAFOs. Wing and Wolf (2000) demonstrated people living closer to intensive pig farms "increased occurrences of headaches, runny nose, sore throat, excessive coughing, diarrhea, and burning eyes" than those living further away

(2000, p. 233). Guo et al. (2021) demonstrated that zoonotic viruses, bacteria, and pathogens (those which can be transmitted from animals to humans), such as E. coli, Streptococcus, Staphylococcus, and Cryptosporidium (some of which display high levels of antibiotic resistance as a result of the antibiotics used on the livestock), are significantly higher closer to CAFOs and urge that "Public health measures should also be implemented to prevent the emergence of new zoonotic pathogens in CAFOs" (Guo et al.. 2021, p. 8).

From a human health perspective, removing pig and cow meat from the human diet could improve the health outcomes for many of those living in high-income countries (Miller et al. 2022), by reducing rates of cancer, cardiovascular disease, coronary heart disease, and diabetes, all of which cost millions of human lives, and significant proportions of healthcare budgets. Since removal of pig and cow meat would also necessitate the closure of CAFOs, the health problems associated with living close to these operations would also be significantly lessened.

At the same time, it is unlikely to have a significant negative impact on health outcomes of those living in countries in the Middle East and North Africa (MENA), sub-Saharan Africa, and South Asia, where excessive red meat consumption is less problematic (Miller et al. 2022), provided alternative sources of protein can be developed on a suitable scale (see below). Such an outcome would contribute significantly to the achievement of SDG 3 (ensure healthy lives and promote well-being for all at all ages).

In addition to the varying health issues associated with the production and consumption of pig and cow meat, there are also several important issues around social inequality. The farming of pigs and cattle, particularly when conducted on an industrial scale, for example, has significant implications here. When land traditionally used or inhabited by Indigenous and local communities is repurposed for agricultural enterprises without adequate compensation or consultation, this can lead to displacement and loss of livelihoods. In turn, this raises issues regarding equity, access to land, and the preservation of cultural heritage, as land dispossession often disproportionately affects already marginalised communities. Scheidel et al (2023) demonstrate that the majority of land conflicts and landscape loss are associated with the "agriculture, forestry, fisheries and livestock (AFFL) sector" and that this undermines the rights of Indigenous peoples (2023, p. 1). By doing so, this undermines the core principles of environmental justice (Schlosberg 2009). In rural China, the dispossession of rural pig farmers by the industrialisation of agriculture leads to an "erosion of agricultural knowledge and practice" (Schneider 2017, p. 90).

Similarly, in Latin America, particularly in Brazil, the expansion of pasturelands for cattle has been a major driver of deforestation in the Amazon rainforest. Hecht and Cockburn (2011) make the point that, while this was done for the notional purpose of using the land more effectively:

> 'Effective use in Amazônia usually resulted in clearing huge areas of forest for cattle production, often displacing forest dwellers of all kinds, from Indians to peasants. (p. 168)

In the summer of 2024, the 7th Federal Environmental and Agrarian Court of the Seção Judiciária do Amazonas in Brazil ruled in a 25-page judgement that a rancher had illegally cleared 5600 ha (56 km^2/21.6 square miles) of the Amazon for cattle farming. The Court ordered that the rancher's assets be frozen to the value of the estimated damage, calculated at R$292 m (£41 m/US$52 m) (IBAMA v Dirceu Kruger). This case only relates to damage caused in relation to climate change and IBAMA (Instituto Brasileiro do Meio Ambiente e dos Recursos Naturais Renováveis—the Brazilian Institute of the Environment and Renewable Natural Resources), who brought the case, say that they reserve the right to being future cases relating to biodiversity damage (FEAC 2024).

This is not, of course, a unique trait of cattle farming, and similar displacements also occurred in the same region in the context of soybean farming (Hecht 2005). For many indigenous and rural communities, the land is not just a resource but a part of their cultural and spiritual identity (Tobvych et al. 2018), and so loss of land, or loss of access to land, may lead to a loss of identity. Ellwanger et al (2020) are among those who draw a link between deforestation of the Amazon and a rise in infectious "zoonotic diseases and vector borne infections" (2020, p. 2).

Addressing the environmental justice impacts of land dispossession requires robust legal and policy frameworks that protect the rights of indigenous and local communities. This includes enforcing land tenure rights, consulting communities before initiating agricultural projects, and compensating those affected by land acquisitions. Internationally, instruments such as the United Nations Declaration on the Rights of Indigenous Peoples (UNDRIP) provide guidelines for the protection of indigenous lands and resources, although adoption and implementation remain inconsistent (UN GA 2017; Vrdjak 2018).

Land dispossession and (effectively) forced resettlement is just one of the social inequalities that may result from pig and cattle farming. CAFOs, the use of which is increasing year on year globally, are rarely situated in urban areas with high land prices. Son et al. (2021) showed that "CAFOs are located disproportionately in communities with higher percentage of minorities and in low-income communities" (2021, p. 2). This disparity between those who reap the benefits of the CAFO (namely the animal-industrial complex and the consumers) and those who beat the brunt of the cost (namely those who are displaced or live near to CAFOs) fits the classic model of environmental injustice as set out by Bullard (1990) and others. The expansion of CAFOs can lead to a loss of access to land for traditional uses such as hunting, fishing, and gathering, deeply affecting these cultural practices. The transformation of landscapes from diverse ecosystems to monocultures dedicated to livestock diminishes not only biodiversity (see below) but also the cultural richness of the area.

As we will see below, the economic need of the animal-industrial complex for "efficient" means of production of pig and cow meat means that the move towards CAFOs will continue. Indeed, in December 2019 the Ministry of Agriculture and Rural Affairs in China announced that it was keen to promote multi-storey indoor pig farms (MARA 2019), and many of these are now in production or competed, including one in Hubei Province which stands 26 storeys high, and will have the capacity to kill over a million pigs annually (Guardian 2022). Small, local farmers often cannot compete with the scale and efficiency of large operations, leading to a decline in traditional farming methods and loss of local jobs, further impacting the economic structure of these communities (Gurian-Sherman 2008, p. 1).

Overall, from an environmental justice perspective, the impacts of pig and beef production and consumption appear to be largely negative. There are many health risks associated with over-eating this type of meat, and with working or living near the CAFOs where much of it is produced. If the trade in pig and cow meat were to cease immediately, then from a strictly environmental justice perspective this would be favourable.

The next sections of the chapter will look at the arguments from an ecological justice and species justice perspective.

Ecological Justice

The section on environmental justice explored the impact of land dispossession caused by pig and cattle farming on the human inhabitants, but it has an equal—and perhaps greater—impact on the non-human inhabitants. This would fall under White's (2022) classification of *ecological* justice.

Space

As identified in the introduction to the book, all foodstuffs require physical space in which to be grown. For pigs and cattle, this space may be, at different times, and in different locations, outside in a field or a woodland, or indoors in a barn, a shed, or a purpose-built multi-storey building. The impacts of these uses of space on specific ecosystems will differ, but this section will draw out the key consequences of pig and cattle farming globally.

As a general rule, land which is used for grazing cattle or raising pigs cannot be used at the same time for forestry and vice versa. The UN Food and Agricultural Organization calculated in 2012 that over a quarter of all the land which is not frozen is used for livestock grazing, and a third of all land used for growing crops is used to grow crops for animal feed (UN FAO 2012). When an area is cleared for animal grazing, it is not just the pasture land itself which is lost. Access roads, buildings, and other farming infrastructure all require additional space. In South America, for example, the UN FAO estimates that "70% of previous forested land in the Amazon is occupied by pastures, and feedcrops cover a large part of the remainder" (2006, p. 23). On a wider scale, Gibbs et al. (2010) used satellite data to calculate that "more than 55% of this new agricultural land came from intact forests" and an "additional 28% ... came from disturbed forests that previously had been affected by shifting cultivation, logging, fuel wood collection, or other forms of gradual degradation" (Gibbs et al. 2010, p. 16733).

The impact of the loss of rainforest cover in terms of climate change has been well documented (for example, Brienan et al. 2015). In the last ten years, the three largest rainforests in the world (the Amazon, Congo Basin, and Australasian) have lost around 30 million hectares of their primary forest cover, most of it to livestock farming (Mongabay 2024).

Not only is the ability of such forest to absorb carbon from the atmosphere being reduced, but as the vegetation is burned, the stored carbon is re-released back into the atmosphere.

What is also a factor is that many species are hunted because of the real, perceived, or imagined threat they pose to farmed livestock. Wolves were hunted nearly to extinction in the United States (Kompaniyets et al. 2017) because of the fear they would kill cattle, badgers are culled in the United Kingdom based on contentious evidence that they transmit tuberculosis to cattle (Cassidy 2019; Cressey 2020), and red foxes are killed because of their predation of piglets in Australia (Fleming et al. 2016). None of these examples happen covertly, they were all undertaken with appropriate government licences. Equally, wildlife may be killed as a result of negligence or accident. Radhakrishnan (2018), for example, found incidents of poisoning of macaques, wild board, gaur (a type of wild cattle), and elephant in Southern India and was able to link all to chemicals and pesticides used in farming. As livestock farming encroaches more and more on previously undisturbed areas, "human-wildlife conflict" levels increase (WWF 2024). This term is problematic for several reasons. It suggests that this is a conflict which is somehow attributable to both sides, rather than the reality of humans encroaching increasingly on territory previously occupied by wildlife—and usually for the purpose of human gains, rather than wildlife gains. In addition, Pooley et al. (2017), suggest that there is more aggravation due to human fear of predators than there is to the damage the predators actually cause—the perception is worse than the reality. Finally, the current way to resolve human-wildlife conflicts is to kill the wildlife, so this is a wholly unequal conflict. It is started by humans, fought against an "enemy" which is causing less damage than feared, and which ends in extermination of the perceived threat.

Even where pigs and cattle are farmed in CAFOs, Wuerther (2010) points out that there are significant biodiversity impacts caused. Son et al (2021), who, as we saw earlier, concluded that pig CAFOs in North Carolina are disproportionately located in low-income and marginalised communities, also confirmed that the animal waste produced can pollute air, water, and soil and thus can have an indirect impact on local biodiversity in those three environmental media.

Place

The farming of cattle for meat production (as opposed to dairy production) generates by far and away the highest greenhouse gas emissions of any foodstuff. Measured per 100 g of protein, beef production emits 49.8 kg CO_2-eq (carbon dioxide equivalent), which is more than double that produced by any other food (Poore and Nemecek 2018). Pig meat, by contrast, produces just 7.61 kg CO_2-eq per 100 g of protein, which is lower than some fruits and vegetables (Poore and Nemecek 2018). In terms of land and water use, beef requires 184 m^2 of land and 728 litres of freshwater per 100 g of protein, and pig meat 10.7 m^2 and 1100 litres of freshwater (Poore and Nemecek 2018). The impact of space requirements for pig farming is lessened in the multi-storey indoor facilities such as those in China mentioned earlier (MARA 2019; Guardian 2022), but there are other environmental costs involved with the construction and operation of such buildings.

With forested areas across the tropical region being cleared for farming, the role played by the vegetation in flood prevention has been significantly eroded. In the Democratic Republic of the Congo, record level floods occurred in the summer of 2023 and early 2024, accompanied by large mudslides. Recognising the link to deforestation, the local authorities in Kalehe territory "launched a campaign to plant 1000 trees on the bare hills" (Nyirabihogo 2024). In Haiti, a combination of urbanisation and deforestation is causing more, and more severe, episodes of flooding (Madoche et al. 2024). It has also been well documented that deforestation is a significant contributor to climate change (Shukla et al. 1990; Nobre et al. 1991 et seq.), and that one of the results of climate change is an increase in extreme weather patterns. Not only that, but in some areas longer periods of flooding reduce the ability of wetland species of tree to reproduce (Greet et al. 2020), so there is a feedback loop created—fewer trees lead to more flooding, which leads to fewer trees, and so on.

The ecological costs of the pig and cow aspects of the animal-industrial complex are not restricted to those around deforestation. Ecologically, the least damaging food to eat is that which you have produced yourself. The idea of being able to trace the origins of food is nothing new—as Thompson (2015) and others have pointed out, it was the norm until the early twentieth century when the industrialisation of food production began to sever the link between producer and consumer. Nonetheless, the concepts of "farm to plate" or "field to fork" have become much more widespread, as has the concept of "food miles," as consumers start to realise the impact of moving foodstuffs across the world. Li et al (2022) report that global food miles account for nearly 20% of all the emissions from the food industry and, since Crippa et al (2021) calculated that food systems are responsible for "a third of global anthropogenic GHG emissions," it stands that food transportation contributes 6% of all anthropogenic greenhouse gas (GHG) emissions. Wiedemann et al (2015) looked specifically at beef and lamb production in Australia where the meat was exported to the United States and confirmed that despite the high ecological cost of food transportation in general, beef and lamb transportation contributed less than 5% of the total emissions attributable to beef and lamb production (Wiedemann et al. 2015). In the UK, cow and pig meats make up around 80% of meat products imported, and about three quarters of the meat consumed (BMPA 2024). The Republic of Ireland provides the vast majority of these imports, limiting the impact of food miles on UK consumed products (BMPA 2024).

Waste

All animals, human and non-human alike, create waste products as a natural consequence of their existence. These waste products may take the form of gases, liquids, or solids. Gases will include both those exhaled like CO_2 and those released as part of food digestion, such as methane (Ch_4). Both of these are known GHGs and contribute significantly to anthropogenic climate change. Liquid waste is primarily urine but also includes blood and other bodily fluids released during slaughter. There appears to be no "standard" measure of cow urination volumes, but figures from Edwards et al. (2015) suggest that it would be between 20.8 and 36 litres per day (2015, p. 81).[2] For pigs, the equivalent figure appears to be around 5 litres per day (Michalak et al. 2020). Given the high levels of nitrogen contained in mammalian urine, this is likely to leach both into and through soil, contaminating both the land itself and the groundwater. Nennich et al (2005) calculated that cattle produce around 65 kg per day of manure (2005, p. 3725), although they do cite earlier estimates of around 56 kg per day (ibid.). Michalak et al (2020) observed that pigs on average excreted about 5 kg of faeces per day.

[2] This figure is the result of my calculations based on Edwards et al.'s data of between 1.8 and 2.4 litres per urination and 11.6–15 urinations per day.

Where the cattle and pigs are raised in CAFOs, the concentration of waste is significantly increased, and the ability of the local environment to simply absorb it is decreased. Mallin (2000) outlines how the liquid and solid waste is piped out of the buildings in which the animals are raised and stored in large lagoons, to be pumped onto fields if the level gets too high (2000, p. 28). The waste problems caused by population concentration are as true of humans as they are of non-human animals, and cities have always struggled more to deal with the volume of with waste produced than rural areas. In the mid-1850s in London, for example, the volume of waste entering the River Thames caused the "Great Stink" of 1858 and caused the then Chancellor of the Exchequer Benjamin Disraeli to describe the river as "a Stygian pool, reeking with ineffable and intolerable horrors" (Disraeli 1858).

It is true that some of the waste from pig and cattle farming can be used as fertiliser—either for crops produced to feed other pigs and cattle, or for crops to feed humans. In the majority of cases where this does not happen, the waste will simply enter the nearest watercourse, either through "non-point-source run-off" (where it is washed from the land into the water by rain) or by draining into the soil and rock until it hits the water table and contaminates the groundwater. This waste often contains pathogens (Gerba and Smith 2005), heavy metals, and nutrients like nitrogen and phosphorus, which can lead to the contamination of water bodies (Burkholder et al. 2007). When these contaminants enter water systems, they can cause *eutrophication*, a process that leads to excessive growth of algae and subsequently decreases the oxygen levels in water, killing fish and other aquatic life. In North Carolina, for example, the river Reuse is hypoxic (very low Oxygen) or anoxic (zero Oxygen) for a 65 km stretch. Mallin et al (2006) attribute run-off from CAFOs and agricultural land as being among the causes of this eutrophication (2006, p. 698). In hypoxic and anoxic rivers, little or no aquatic life survives, and this has impacts for all species, including humans, who are connected to this ecosystem.

What we can see is that the ecological damage caused by cattle and pig farming—particularly in CAFOs—is significant. This may be from deforestation, the run-off of pollution into watercourses, or the harder-to-account-for impacts of transportation of both feedstuff and the livestock or meat itself. The next part of the chapter will assess the species justice aspects of this farming, focusing on the biodiversity costs.

Species Justice

This section of the chapter will discuss the welfare of cattle and pigs raised for human consumption, and also the impact of farming on the wider biodiversity. The specific biodiversity impacts will vary, depending upon whether we are looking at the areas around CAFOs in North Carolina, multi-storey pig "farms" on China, or outdoor areas of cleared rainforest in Africa and South America. First though, we will look at the issue of cattle and pig welfare.

Most studies which consider the welfare of non-human animals focus on those which are sentient, that is to say those which have "the capacity to experience one or more of the various states we call feelings" (Broom 2019). Indeed, even though Mullan et al (2024), who propose the creation of a "sentiocentric utilitarian framework" to decide which species could be farmed, do not specifically exclude non-sentient species from their calculations, they make it clear that they feel "non-sentient non-typical species are likely to be most acceptable to farm" (2024, p. 13). Szûcs et al (2012) considered the issue of animal welfare from different cultural and religious perspectives and concluded that "most people think that extremes of poor welfare in animals are unacceptable and that those who keep animals should strive for good welfare" (2012, p. 1499).

Both cattle (Wilkie 2010) and pigs (Garcia and McGlone 2022) are recognised as sentient vertebrates and, as such, are protected by a range of welfare legislation—from "Martin's Act" (the Cruel Treatment of Cattle Act 1822) which was the first specific animal welfare legislation in the world, to the present day. Every piece of this legislation has limits and boundaries which allow for certain actions which might otherwise been considered "wrong" to fall within what is permissible. White (2022, p. 19) distinguishes between illegal harms (those prohibited by criminal law) and legal harms (those not prohibited). For example, not only it is not an offence in the UK to puncture the ear of a newborn calf in order to attach an eartag, but it is a criminal offence not to do so (r2(4) The Cattle Identification Regulations 1998). This would be a "legal harm." To do the same to a kitten would be illegal, under the Animal Welfare Act 2006, and this is an "illegal harm."

Where there are regulations promoting animal welfare (rather than banning animal cruelty), they are largely based on the "Five Freedoms" of Animal Welfare. These emerged from the Brambell Committee (the Technical Committee to Inquire into the Welfare of Animals Kept under

Intensive Livestock Husbandry Systems) which was itself a response to Harrison's "Animal Machines" (Harrison 1964). The original "freedoms" outlined in the Brambell Report were that:

> An animal should at least have sufficient freedom of movement to be able without difficulty, to turn round, groom itself, get up, lie down and stretch its limbs. (Brambell 1965, Para 37)

The freedoms all related to movement, though it should be noted that Brambell also made suggestions around feeding, conditions for housing, and awareness of behavioural issues. These wider ideas were developed by the UK Farm Animal Advisory Committee (subsequently Farm Animal Welfare Council) into:

> freedom from hunger, thirst or malnutrition; appropriate comfort and shelter; prevention, or rapid diagnosis and treatment, of injury and disease; freedom to display most normal patterns of behaviour; and freedom from fear. (FAWC 2009, p. 2)

The concept has evolved again since the FAWC report, and the versions most currently used today are:

- Freedom from Hunger and Thirst
- Freedom from Discomfort
- Freedom from Pain, Injury or Disease
- Freedom to Express Normal Behaviour
- Freedom from Fear and Distress.

These are reflected in the Animal Welfare Act 2006 in England and Wales, and across Europe through either the Council of Europe's 1976 Convention for the Protection of Animals kept for Farming Purposes, or in European Union under the 2007 Treaty of Lisbon (Article 13 of Title II), or both. In the United States, the Animal Welfare Act of 1966 is the primary piece of federal Legislation and specifically excludes farm animals (§2132(g)(3)). The US National Agricultural Law Centre says that at "the state level, laws to prevent deliberate animal cruelty sometimes apply to farm animals, but few states have prescribed on-farm treatment standards" (NALC 2024). China has no legislation covering the welfare of farmed animals, and in India, the Prevention of Cruelty to Animals Act

1960 (as amended) imposes a duty on "every person having the care or charge of any animal to take all reasonable measures to ensure the well-being of such animal and to prevent the infliction upon such animal of unnecessary pain or suffering" (s3). The Indian legislation is wider even than it first appears, since "animal" is defined in s2(a) as "any living creature other than a human being," thus obviating the requirement for sentience to be proven.

Generally, improving animal welfare standards will involve additional expense for the farmer. If cattle are required to have twice as much space, then half as many can be kept in the same building, so either the profits must fall, or the price to the consumer must rise. This triggered Stamp Dawkins to state that the "the case [for improvements in welfare] will have to be particularly clear and watertight, because every excuse will be found to turn it down" (2016, p. 4).

The existence of a national legislative framework to prevent animal cruelty and promote animal welfare is a solid starting point but in this area of law, as in all others, it is predicated on a strong enforcement system, and is largely reactive—the animal has to suffer and be seen to suffer before any action can be taken.

Even when animals are shown to suffer as a result of farming practices, the outcome depends on location. Prunier et al (2010) have demonstrated that "High physiological demands of modern types of pigs in intensive piggeries [CAFOs] are likely to induce health and welfare problems" (2010, p. 895), and … says of cattle in CAFOs that they are "standing ankle-deep in their own waste eating a diet that makes then sick" (p. 317) and yet the use of CAFOs is increasing rather than ceasing. Wyatt correctly identifies that "the abuse that takes place during factory pig farming can be characterised as continual and pervasive yet is largely within the legal sphere" (p. 15) which leads us back to White's distinction of legal and illegal harms (2014, p. 19). In addition, Thompson (2015) makes the point that animals in CAFOs "are fed on grains from industrial monoculture" (p. 187) thus spreading the environmental dystopia beyond the animals themselves.

The commercial rearing of cattle and pigs is not guaranteed to meet welfare standards (where they exist). This allows us to bypass Zangwill's (2021) argument raised in Chapter 1 painting the consumption of meat as "mutually beneficial" since his argument is predicated on farm animals which live "good lives," something which those raised in CAFOs patently do not. From a species justice perspective then, a sudden end to the commercial farming of cattle and pigs would have a beneficial impact. We now need to turn to the wider biodiversity impacts of such farming.

The most significant impact that beef farming has on biodiversity is when it involves deforestation, particularly in areas such as the Amazon Rainforest and the Congo basin. The IUCN has found that "forest-specialist species are on average more threatened than the migratory species and show a steeper decline towards extinction" (IUCN 2024). Species, whether endangered or not, are not uniformly distributed across habitats, and so it is impossible to pluck a number of endangered species per hectare of rainforest with any accuracy. However, a ground-breaking report by the Science Panel for the Amazon in 2021, estimates that "single plots of one hectare may have over 300 tree species" (2021, p. 10). It is important to recognise that the IUCN Red List contains over 850 Critically Endangered species in South America, and close to 1000 in sub-Saharan Africa, which are living in forests threatened by agriculture (IUCN 2024), including mammals, amphibians, reptiles, and plants. Some, for example the Sehuencas Water Frog (*Telmatobius yuracare*), are estimated to number under fifty, but the remoteness of their habitat makes accurate assessment difficult.

Pig farming is much less likely to involve the clearance of rainforests and is more likely to take place in CAFO type surroundings. Ruckli et al. (2021) looked at a range of different type of pig farm and concluded "regardless of farm type, scores for ecosystem, species and genetic diversity were scored on a low level so that the overall biodiversity theme score was between 38 and 43%" (2021, p. 14). Bartlett et al. (2023) concur that in order to improve biodiversity scores around pig farms, there needs to be a trade off with the stocking density—largely because of the increased antimicrobial use in CAFOs than in have much lower biodiversity scores than larger, less dense farms. Ironically, this suggests that while the multi-storey CAFO of the type used in China (MARA 2019; Guardian 2022) is bad for the health of the pigs and surrounding inhabitants, it does involve lower areas of land use than a single-storey CAFO with the same stock numbers, so it might actually be less damaging for biodiversity.

Having considered the environmental, economic, and species justice impacts of current pig and cattle farming practices, the vast majority of which are negative, we will not turn to consider the economic impact and consider whether the financial rewards for such a trade outweigh the costs.

Economics

Market calculation for pig and cattle meat is calculated in "carcass weight equivalent" (cwe) which allows for some standardisation of the wildly differing cuts of meat which are traded. Eurostat, the statistical office of the European Union, defines the carcass weight of pigs and cattle as:

- "For pigs it is the weight of the slaughtered pig's cold body, either whole or divided in half along the mid-line, after being bled and eviscerated and after removal of the tongue, bristles, hooves, genitalia, flare fat, kidneys, and diaphragm.
- For cattle it is the weight of the slaughtered animal's cold body after being skinned, bled, and eviscerated, and after removal of the external genitalia, the limbs, the head, the tail, the kidneys and kidney fats, and the udder." (Eurostat 2024)

The global beef market for 2024 is largely unchanged from that of 2022 and is around 60 million metric tons cwe, of which around 10 million tons cwe are exported from the country of production (USDA 2024a). For pig meat, the equivalent figure is around 115 mtcwe, of which approximately 10 million tons cwe is exported (USDA 2024a, b). In financial terms, the global beef market size is projected to grow to US$ 60bn by the end of 2024 (FBI 2024), which is larger than the global pig meat market of US$250bn (RAM 2024). The combined market is thus around US$700bn a year, which puts it at the same level as the GDP of a relatively prosperous nation (the IMF estimate of Belgium's GDP as US$655bn for 2024) (IMF 2024).

The market for the meat products is, of course, only a fraction of the total economic impact of the industry. Cattle farm workers, cattle farm owners, CAFO building companies, feedstock producers and growers, transportation companies, slaughterhouses, butchers, retailers, cooks, and chefs—all have a financial relationship with the meat industry. This makes the true value of the industry unquantifiable, but undeniably significant. If the pork and beef industries were to suddenly cease to exist, the financial implications would be significant. This section of the chapter will look at whether this impact would be equally distributed, or whether some communities are more reliant on others on the pork and beef industries.

The countries which produce the most beef in the world are, in order, the United States, Brazil, and China who between them produce just over half of the beef globally (USDA 2024b). The top three pork-producing states are, also in order, China, the United States, and Brazil, who between them produce almost two-thirds of the pork globally (USDA 2024b).[3]

In the United States, Texas, Oklahoma, and Missouri raise almost 8 million cattle for beef production, which is more than a quarter of the total for the whole country (Cook 2024). Iowa produces the largest number of pigs (Pork 2024), although the largest farms are in North Carolina, which is the largest corporate pork producer in America being owned by a Chinese company, WH Group (Thomas 2023). In terms of per capita personal income ratings, Texas was 24th highest, Iowa was 33rd, Missouri was 35th, and Oklahoma was 43rd, so none of the top pork- and beef-producing states sits in the top 20 wealthiest states in the country (EDA 2024).

In China, beef production has risen to historically high levels, although the country is still an importer of beef, primarily from Brazil (Smith 2023). Prices for beef fell dramatically from 2023, and are currently about half of the traditional levels (Nulimaimaiti and Huifeng 2024). Henan Province produces the most beef (AHDB 2020) and has the 8th lowest level of disposable income per capital in China, at around 25% lower than the national average (NBS 2024). Sichuan, Henan, and Hunan provinces produce more pork than any others (Zhang et al. 2021), and all three are below the Chinese average for disposable income per capita (NBS 2024).

Mato Grosso is the Brazilian state which produces the most beef (Yoshikawa 2023), and Santa Carolina is the largest pork-producing state (Savian et al. 2024). Mato Grosso is sparsely populated and includes part of the Amazon rainforest within its boundaries and while it is relatively wealthy overall (the third wealthiest state in the country) "poverty and inequality [...] persist" (World Bank 2024). Santa Catarina too is wealthy, coming in 4th in terms of GDP per capita (GdSC 2022).

Pig farming and cattle farming are clearly lucrative businesses, and the national impact on wealth of any changes to the market is finely interwoven—the largest pig farming company in the United States is Chinese owned, China is a significant import market for Brazilian beef, and so on.

What has not been factored in to these figures are the criminal enterprises which are engaged in both pig farming and cattle farming. Kington

[3] The European Union is classed by the USDA as the third largest pork producer but has been discounted here for not being a single country.

(2024) reports that in Puglia, southern Italy, a local group within the Sacra Corona Unita (the regional Organised Crime Group) have started to reinvest laundered money in cattle farming, because of the high levels of subsidy which can be gained. This is not a new phenomenon, and Tondo reported in 2017 that La Cosa Nostra was doing something similar in Sicily (Tondo 2017), it is not just a way of laundering money, and cattle are valuable commodities and thus worth stealing (Olaniyan and Yahaya 2016) or cutting down the Amazon (Sesnie et al. 2017). Organised Crime Groups seem to be less involved in pig farming, although there is a notable rise in livestock theft in the UK (Tudor 2024).

The removal of pork and beef from the human food chain will have a significant financial impact across the globe. In the United States and China, this impact may be felt predominantly in poorer regions, as the main cattle- and pig-producing regions are largely below average in terms of wealth. In Brazil, the initial results suggest that Mato Grosso and Santa Catarina are among the wealthiest parts of the country, but the wide disparity in income between the rich and poor is obscuring the true results. In the United States, where the vast majority of cattle and pig farms are family owned, the distribution of subsidies means that very large, corporate-owned farms "with sales of $5 m or more – accounted for fewer than 1% of all farms but 42% of all sales" (Lakhami 2024). The Brazilian company JBS is the largest beef and pork processing company in the world and, in addition to controlling the industry in Brazil, is the second largest producer in the United States (JBS 2024). Unpicking ownership in China is harder, as Chinese law forbids private ownership of rural land, putting it into the hands of a village collective (Chen and Jin 2023). Hong Kong-based WH Group,[4] mentioned above, is "the largest pork company on the world" (WH Group 2024), a title for which it competes with JBS. It operates globally and generated year-on-year revenue of US$6.1bn in April 2024.

The small number of very large multinational players who control the beef and pork markets means that the economic impacts, if these industries were to cease, would be large scale and far reaching. The power of the animal-industrial complex to protect its own interests is also significant. The next section of the chapter looks at possible future directions for the industry and alternatives to pork and beef farming. Some of these are

[4] Sometimes referred to as Shuanghui Group (the group's previous name) or Shineway Group.

largely theoretical, some are actively being developed, and some already exist.

Future and Conclusion

This final part of the chapter will consider the legal and social challenges which would need to be overcome if the beef and pork markets were to disappear, as well as signposting the discussions in the Conclusion chapter. If the Covid-19 pandemic has shown us anything, it is that governments of all political persuasions have the wherewithal to close entire industries if it is deemed to be in the public interest. Bars, hotels, hairdressers, cinemas, schools, and airlines were all effectively closed in different countries because of lockdown restrictions, brought about because of a stated interest in protecting human health. Whatever the reason, the end of pig and cattle farming would require an immediate and significant shift to alternative sources of protein, so that is where we will go first.

The World Economic Forum identified in 2022 that while across Europe, the United States, and many other regions of the developed world, more and more people are adopting a vegetarian or vegan diet, the rates are falling in countries like China and India (Buchholz 2022). The Independent Panel of Experts on Sustainable Good Systems (IPES-Food) was clear in their 2022 report that "the status quo in animal production systems is not an option, whether we are most concerned with climate change, biodiversity loss, livelihood risks, food security, or animal welfare" (iPES-Food 2022a, p. 15). However, if beef and pork were no longer available, there are three main directions in which consumers could go:

1. A species shift. Moving from pork and beef to sheep, goat, poultry, or other animals.
2. A meat-free diet.
3. The adoption of novel foods, such as lab-grown meat, algae, or insects.

In reality, it is likely that a mixture of all three of these approaches would be taken, so we need to look at all three in the context of the factors we have looked at in the chapter so far—environmental justice, ecological justice, species justice, and economics.

The human health implications, both positive and negative, of shifting consumption from one type of red meat to another (i.e., from cow to goat) will be minimal, so further discussion here is not needed. The benefits and pitfalls of a meat-free diet are more complex. The reduction in saturated fats, increase in unsaturated fats, and the availability in vitamins and minerals (as discussed above) that are offered by a meat-free diet are clearly positives. Trace amino acids such as creatine and carnosine may be harder to replace, but not impossible. McDermott and Wyatt (2017) note that "the proteins in pulses and cereals are complementary, meaning that together they provide adequate amounts of all the essential amino acids" (p. 34). Arsenault and Brown (2017) say that dietary protein deficiencies in children in low-income countries is not a limiting factor, and this suggests that "while protein deficiencies are a reality in specific populations/regions, there is not a global protein deficit" (iPES-Food 2022a, p. 24). One of the important negatives of a meat-free diet is said to be the presents of ultra-processed foods. Cutroneo et al (2022) looked at a range of meat alternatives and found that "all these products are classified as 'ultra-processed' foods" (2022, p. 10). Later research by de las Heras-Delgado et al. (2023) suggested a lower, but still significant figure of 41% (2023, p. 6).

It is not just the health of consumers which needs to be considered. In Britain, for example, 27 people were killed in 2022/2023 due to agriculture-related activities and farming. Eight of those were killed by livestock, primarily cattle, and the rest were killed by machinery, vehicles, or buildings (HSE 2023). None was killed as a result of arable or vegetable farming. As was demonstrated above, those living near to CAFOs are more likely to display a range of health conditions, and the overuse of antibiotics has led to the emergence of antibiotic-resistant infections such as MRSA and e-Coli.

There is some disagreement about whether to classify laboratory-grown meat as processed or ultra-processed. It is clearly not a naturally produced animal, as it has been cultured in a laboratory. It is also possible to engineer such products to contain additional vitamins and minerals, but the industry is still in its infancy and is constantly developing.

Cherry et al (2019) suggest that eating seaweed, which is one possibility to replace meat consumption, exposes the consumer to a risk of iodine and arsenic ingestion, but may also alleviate the risk factors associated with obesity, type-3 diabetes, and cardiovascular disease. A review of previous studies by Cunha et al. (2023) revealed some health benefits of consuming insects as a novel food but were clear that risks such as allergic reactions needed more research.

One of the traditional limitations of livestock farming has been space. Poore and Nemecek (2018) show that in order to produce 1000 kCal of beef or lamb, well in excess of 100 m^2 of land is needed. Pigs and Poultry require around 7 m^2, whereas the most space needed for a non-animal product is the 4 m^2 for tomatoes. The indoor, multi-storey CAFO of the type being used in China goes against this trend, but still creates the issues of waste treatment and the health problems that are associated with it. Ritchie (2024) has used the Poore and Nemecek data to calculate that if the world went to an entirely meat-free diet, then agricultural land-use requirements would drop by 75%. With so much less space needed for agriculture, the justification for deforestation in the Amazon Congo Basin and elsewhere would no longer be necessary, and the environmental impacts of a meat-free consumption pattern could be significant for reforestation and biodiversity gain.

In terms of lab-grown meat, the process requires a starter cell sample to be taken from an animal, but from that point onwards, it is produced in a laboratory. Space, place, and waste are not significant issues to date, though it is worth noting that the industry is still in its infancy. In 2011, Tuomisto et al. calculated that lab-grown meat creates "78–96% lower GHG emissions, 99% lower land use, and 82–96% lower water use" (2011, p. 6117), although more recent analysis (Lynch and Pierrehumbert 2019) injects a note of caution, concluding that the "climate impacts of cultured meat production will depend on what level of decarbonized energy generation can be achieved, and the specific environmental footprints of production" (2019, p. 10).

From a species justice perspective, the move away from meat consumption is undeniably the preferable approach. A shift from a CAFO containing pigs to one containing sheep or goats would merely transfer the burden from one sentient species to another, and so in terms of welfare there would be no net gain. However, lab-grown meat, insects, and algae are not classed as sentient, so issues of animal welfare would cease to apply if these were to become the replacement foodstuff for pork and beef.

Similarly, a shift to a plat-based diet, with the 75% reduction in land space requirements identified earlier, would bring with it a real opportunity for ravaged areas across the world to be reforested and actively protected. This would create an environment in which species at risk of extinction could find safe harbour and reduce the anthropocentric impact on biodiversity loss through both habitat loss and climate change.

The animal-industrial complex has demonstrated that it is determined not to be sidelined by any shifts in consumption patterns. JBS, for example, is already heavily engaged in the production and processing of poultry, sheep, and goats (and salmon, see Chapter 3) (JBS 2024), so it would be able to adapt its portfolio to replace pork and beef production. The WH Group is almost entirely focused on the pork market and associated businesses, so it would not be able to adapt so quickly. The fact that consumption of lamb is growing in popularity in China may soften the blow of any shift (Wang et al. 2022).

If the shift is to plant-based diets, then as we have seen the production of crops such as soy would need to increase. Mato Grosso already accounts for a significant proportion of Brazil's soybean production, and it would be straightforward to shift production from soybeans grown for cattle feed to human consumption. The same is true of Sichuan and Henan provinces in China and Texas, Iowa Missouri, and Oklahoma in the United states. The only region that is a major producer of pork or beef but not a major producer of soybean is Hunan Province in China, so it is apparent that most regions will be able to adapt. The majority of the large meat-processing companies have also moved to sell meat-free versions of their products, including JBS, WH Group, Marfrig (the second largest meat company in Brazil), and US company Tyson Foods (Coyne 2024).

The third option, a growth in "novel foods" is also an area where the corporate farmers have an interest. In 2023, JBS announced that it was constructing the world's largest "cultivated beef protein plant" in Spain (Grylls 2023). The global independent thinktank iPES-Food has forecast that the "meat substitute" market will have grown to US$28bn by 2025 (iPES-Food 2022b), so while this is not currently forecast to eclipse more established meat production, it is a significant market.

So, where does that leave us? The chapter set out to explore the impacts of a (fictitious) sudden end to pork and beef farming and consumption. We have looked at how this would impact on human health, the wider environment, different species and local, regional, and international economics. In addition, we have considered what might replace this well-established and industrialised industry. What we have found is that the industry is controlled by a small number of very large, multinational corporations (the animal-industrial complex) and that this complex is simultaneously pushing the message that meat is essential to human health and well-being and investing heavily in alternatives. This is the same approach that is being taken in terms of fossil fuels, where multinational oil corporations and oil-rich states have downplayed and disputed the impact of fossil fuels on the climate for decades yet are the largest investors in renewable energy. The physical health impacts to humans of consuming less red meat are no longer debatable. Equally, the benefit to cows and pigs of not living a life in a CAFO which is nasty, brutish, and short is unequivocal. Environmentally, moving away from beef and pork consumption results in lower GHG emissions, lower land use, lower water consumption, less risk of antibiotic-resistant pathogens, and a rise in farmworker safety. Economically, the areas which produce the most pork and beef are already well established to shift to crop-based proteins, so it will not feel a financial burden.

Taking all of these factors into account, we have to conclude that it is net beneficial to end pork and beef consumption and, since that is not feasible given the entrenched nature of the industry, to dramatically reduce it.

Sources

Arsenault, J.E., and K.H. Brown. 2017. Dietary protein intake in young children in selected low-income countries is generally adequate in relation to estimated requirements for healthy children, except when complementary food intake is low. *The Journal of Nutrition* 147 (5): 932–939. https://doi.org/10.3945/jn.116.239657.

Baguet, A., J. Bourgois, L. Vanhee, E. Achten, and W. Derave. 2010. Important role of muscle carnosine in rowing performance. *Journal of Applied Physiology* 109 (4): 1096–1101. https://doi.org/10.1152/japplphysiol.00141.2010.

Bartlett, H., A. Balmford, J.L.N. Wood, and M.A. Holmes. 2023. Identifying ways of producing pigs more sustainably: Tradeoffs and co-benefits in land and antimicrobial use. *Science and Reports* 13 (1): 2840. https://doi.org/10.1038/s41598-023-29480-5.

Bekdash, R.A. 2023. Methyl donors, epigenetic alterations, and brain health: Understanding the connection. *International Journal of Molecular Sciences* 24 (3): 2346. https://doi.org/10.3390/ijms24032346

Blancquaert, L., A. Baguet, T. Bex, A. Volkaert, I. Everaert, J. Delanghe, M. Petrovic, C. Vervaet, S. De Henauw, D. Constantin-Teodosiu, P. Greenhaff, and W. Derave. 2018. Changing to a vegetarian diet reduces the body creatine pool in omnivorous women, but appears not to affect carnitine and carnosine homeostasis: a randomised trial, *The British Journal of Nutrition; Cambridge* 119 (7): 759–770. https://doi.org/10.1017/S000711451800017X.

BMPA. 2024. Imports and exports. British Meat Processors Association. https://britishmeatindustry.org/industry/imports-exports/.

Bouvard, V., D. Loomis, K.Z. Guyton, Y. Grosse, F. El Ghissassi, L. Benbrahim-Tallaa, N. Guha, H. Mattock, and K. Straif. 2015. Carcinogenicity of consumption of red and processed meat. *The Lancet Oncology* 16 (16): 1599–1600. https://doi.org/10.1016/S1470-2045(15)00444-1.

Brambell. 1965. *The report of the technical committee to inquire into the welfare of animals kept under intensive livestock husbandry systems*. London: HMSO.

Brienan, et al. 2015. Long-term decline of the Amazon carbon sink. *Nature* 519 (7543): 344–348. https://doi.org/10.1038/nature14283.

Broom, D.M. 2019. The concept of sentience. In *Encyclopedia of animal behaviour*, 2nd ed., ed. by J.C. Cho. London: Academic Press.

Buchholz, K. 2022. The rise (or fall?) of vegetarianism. World Economic Forum. https://www.weforum.org/agenda/2022/11/vegetarianism-rise-fall-world-chart.

Bullard, R. 1990. *Dumping in Dixie: Race, class and environmental quality*. New York: Routledge.

Burkholder, J., B. Libra, P. Weyer, S. Heathcote, D. Kolpin, P.S. Thorne, and M. Wichman. 2007. Impacts of waste from concentrated animal feeding operations on water quality. *Environmental Health Perspectives* 115 (2): 308–312. https://doi.org/10.1289/ehp.8839.

Cassidy, A., ed. 2019. *Vermin, victims and disease: British debates over bovine tuberculosis and badgers*. London: Palgrave Macmillan.

Chen, H., and AHDB. 2020. The Beef Market in China. Agriculture and Horticulture Development Board. https://ahdb.org.uk/.

Chen, Y., and Jin, X. 2023. How do farmers realize their rights on the collective land in rural China? An explanatory framework for deconstructing the subject of collective land ownership. *Land* 12 (9): 1746. https://doi.org/10.3390/land12091746.

Cherry, P., C. O'Hara, P.J. Magee, E.M. McSorley, and P.J. Allsopp. 2019. Risks and benefits of consuming edible seaweeds. *Nutrition Reviews* 77 (5): 307–329. https://doi.org/10.1093/nutrit/nuy066.

Clifton, P.M., and J.B. Keogh. 2017. A systematic review of the effect of dietary saturated and polyunsaturated fat on heart disease. *Nutrition, Metabolism, and Cardiovascular Diseases* 27 (12): 1060–1080. https://doi.org/10.1016/j.numecd.2017.10.010.

Collinge, J. 2001. Prion diseases of humans and animals: Their causes and molecular basis. *Annual Review of Neuroscience* 24: 519–550. https://doi.org/10.1146/annurev.neuro.24.1.519.

Cook, R. 2024. Ranking of states with the most beef cows. National beef Wire, 6 February. https://www.nationalbeefwire.com/ranking-of-states-with-the-most-beef-cows-texas-stands-as-the-undisputed-leader-boasting-a-staggering-4-115-million-head-of-beef-cows.

Coyne, A. 2024. Eyeing alternatives—Meat companies with stakes in meat-free and cell-based meat. Just Food, 11 April. https://www.just-food.com/features/eyeing-alternatives-meat-companies-with-stakes-in-meat-free-and-cell-based-meat/.

Cressey, D. 2020. UK official defends badger cull. *Nature*. https://doi.org/10.1038/nature.2013.13136.

Crippa, M., E. Solazzo, D. Guizzardi, F. Monforti-Ferrario, F.N. Tubiello, and A. Leip. 2021. Food systems are responsible for a third of global anthropogenic GHG emissions. *Nature Food* 2: 198–209. https://doi.org/10.1038/s43016-021-00225-9.

Cunha, N., V. Andrade, P. Ruivo, and P. Pinto. 2023. Effects of insect consumption on human health: A systematic review of human studies. *Nutrients* 15 (14): 3076. https://doi.org/10.3390/nu15143076.

Cutroneo, S., D. Angelino, T. Tedeschi, N. Pellegrini, and D. Martini. 2022. Nutritional quality of meat analogues: Results from the food labelling of Italian products (FLIP) project. *Frontiers in Nutrition* 9: Article 852831. https://doi.org/10.3389/fnut.2022.852831.

Dawkins, M.S. 2016. *Why animals matter: Animal consciousness, animal welfare, and human well-being*. Oxford: OUP.

de las Heras-Delgado, S., Shyam, S., Cunillera, E., Dragusan, N., Salas-Salvado, J., and Babio, N. 2023. Are plant-based alternatives healthier? A two-dimensional evaluation from nutritional and processing standpoints. *Food Research International* 169: 112857. https://doi.org/10.1016/j.foodres.2023.112857.

De Otto, M.C.O., A. Alonso, D.-H. Lee, G.L. Delclos, A.G. Bertoni, Rui Jiang, J.A. Lima, E. Symanski, D.R. Jacobs, and J.A. Nettleton. 2012. Dietary intakes of Zinc and Heme iron from red meat, but not from other sources are associated with greater risk of metabolic syndrome and cardiovascular disease. *The Journal of Nutrition* 142 (3): 526–533. https://doi.org/10.3945/jn.111.14978.

Disraeli, B. 1858. Metropolis local management act amendment bill. First Reading, Hansard, HC Deb 15 July, vol. 151, c1508. https://api.parliament.uk/historic-hansard/commons/1858/jul/15/first-reading.

EDA. 2024. 2023 per capita personal income by state, StatsAmerica. US Economic Development Administration/Indiana Business Research Centre, Indiana University. https://www.statsamerica.org/sip/rank_list.aspx.

Edwards, GR., Bryant, RH. Smith, N. Hague, H. Taylor, S. Ferris, A. and Farrell, L. 2015. Milk production and urination behaviour of dairy cows grazing diverse and simple pastures Proceedings of the New Zealand Society of Animal Production. Vol 75: 79 p81.

Ellwanger, J.H., B. Kulmann-Leal, V.L. Kaminski, J.M. Valverde-Villegas, A.B.G. De Viega, F.R. Spolki, P.M. Fearnside, L. Caesar, L.L. Giatti, G.L. Wallau, S.E.M. Ameida, M.R. Borba, V.P. Da Hora, and J.A.B. Chies. 2020. Beyond diversity loss and climate change: Impacts of Amazon deforestation on infectious diseases and public health. *Anais Da Academia Brasileira De Ciências [Annals of the Brazilian Academy of Sciences]* 92 (1): e20191375. https://doi.org/10.1590/0001-3765202020191375.

Eurostat. 2024. Glossary: Carcass weight. https://ec.europa.eu/eurostat/statistics-explained.

FAO. 2024. *Meat market review: Overview of global market developments in 2023.* Rome: United National Food and Agriculture Organization.

FAWC. 2009. *Farm animal welfare in Great Britain: Past, present and future, farm animal welfare council.* London: FAWC.

FBI. 2024. Beef market size, share & industry analysis, by cut type (ground, roasts, steaks, and others), and distribution channel (retail sales, HoReCa, and butcher shops), and regional forecast, 2024–2032. Fortune Business Insights. https://www.fortunebusinessinsights.com/beef-market-106640.

FEAC. 2024. IBAMA v Dirceu Kruger, PROCESSO: 1037196-19.2023.4.01.3200, Seção Judiciária do Amazonas, 7ª Vara Federal Ambiental e Agrária da SJAM. https://www.gov.br/agu/pt-br/comunicacao/noticias/agu-obtem-bloqueio-de-r-292-milhoes-em-bens-de-infrator-ambiental-por-emissao-de-gases-do-efeito-estufa/Deciso_Danos_Climticos.pdf.

Fiego, D.P.L., A.M. Belmonte, and F. Mezzetti. 2018. Pig meat health: A possible binomial? Archivos de zootecnia, PROCEEDINGS IX Simposio Internacional sobre el Cerdo Mediterráneo, 141–145, https://doi.org/10.21071/az.v67iSupplement.3591.

Fleming, P.A., S.J. Dundas, Y.Y.W. Lau, and J.R. Pluske. 2016. Predation by red foxes (Vulpes vulpes) at an outdoor piggery. *Animals* 6 (10): 60. https://doi.org/10.3390/ani6100060.

Garcia, A., and J.J. McGlone. 2022. Animal welfare and the acknowledgment of cultural differences. *Animals* 12: 474. https://doi.org/10.3390/ani12040474.

GdSC. 2022. Santa Catarina: General balance sheet 2022. Goberno de Santa Catarina. https://www.sef.sc.gov.br/api-portal/Documento/ver/4606

Gerba, C.P., and J.E. Smith. 2005. Sources of pathogenic microorganisms and their fate during land application of wastes. *Journal of Environmental Quality* 34 (1): 42–48. https://doi.org/10.2134/jeq2005.0042a.

Gibbs, H.K., A.S. Ruesch, F. Achard, M.K. Clayton, P. Holmgren, N. Ramankutty, and J.A. Foley. 2010. Tropical forests were the primary sources of new agricultural land in the 1980s and 1990s. *PNAS* 107 (38). https://doi.org/10.1073/pnas.0910275107.

Greet, J., S. Fischer, and K. Russell. 2020. Longer duration flooding reduces the growth and sexual reproductive efforts of a keystone wetland tree species. *Wetlands Ecology and Management* 28 (4): 655–666. https://doi.org/10.1007/s11273-020-09738-9.

Grylls, B. 2023. JBS begins construction of world's largest cultivated beef protein plant. Food Manufacture. https://www.foodmanufacture.co.uk/Article/2023/06/09/jbs-building-world-s-largest-cultivated-beef-protein-plant.

Guardian. 2022. China's 26-storey pig skyscraper ready to slaughter 1 million pigs a year. The Guardian, 25 November. https://www.theguardian.com/environment/2022/nov/25/chinas-26-storey-pig-skyscraper-ready-to-produce-1-million-pigs-a-year.

Guo, Y., U. Ryan, Y. Feng, and L. Xiao. 2021. Association of common zoonotic pathogens with concentrated animal feeding operations. *Frontiers in Microbiology* 12: 810142. https://doi.org/10.3389/fmicb.2021.810142.

Gurian-Sherman, D. 2008. CAFOs uncovered: The untold costs of confined animal feeding operations, Union of concerned scientists. https://www.ucsusa.org/sites/default/files/2019-10/cafos-uncovered-executive-summary.pdf.

Harrison, R. 1964. *Animal machines: The new factory farming industry*. London: Vincent Stuart Ltd.

Hecht, S., and A. Cockburn. 2011. *The fate of the forest: Developers, destroyers, and defenders of the Amazon*. Chicago: University of Chicago Press.

Hecht, S.B. 2005. Soybeans, development and conservation on the Amazon frontier. *Development and Change* 36 (2): 375–404. https://doi.org/10.1111/j.0012-155X.2005.00415.x.

HSE. 2023. Summary of fatal injuries in agriculture, forestry and fishing in Great Britain. Health and Safety Executive. https://www.hse.gov.uk/agriculture/assets/docs/agriculture-fatal-injuries-2023-summary.pdf.

Hu, B., X. He, H. Sun, Y. Hu, F. Li, Y. Sun, J. Sun, and L. Feng. 2024. Red and processed meat intake and risk of cardiovascular disease: A two-sample Mendelian randomization study. *Clinical Nutrition ESPEN* 60: 289–297. https://doi.org/10.1016/j.clnesp.2024.02.014.

Hunt, J.R. 2003. Bioavailability of iron, zinc, and other trace minerals from vegetarian diets. *American Journal of Clinical Nutrition* 78 (3 Suppl): 633S-639S. https://doi.org/10.1093/ajcn/78.3.633S.

IMF. 2024. World economic outlook database. International Monetary Fund. https://www.imf.org/en/Publications/WEO/weo-database/2024/April/weo-report.

iPES-Food. 2022a. The politics of protein: examining claims about livestock, fish, 'alternative proteins' and sustainability. www.ipes-food.org.

iPES-Food. 2022b. Fake meat spotlight. International Panel of Experts on Sustainable Good Systems. https://www.ipes-food.org/_img/upload/files/FakeMeatSpotlight.pdf.

IUCN. 2024. Red list index (RLI): The IUCN red list of threatened species. Version 2024-1. https://www.iucnredlist.org.

JBS. 2024. Our locations. JBS. https://jbsfoodsgroup.com/our-locations.

Kington, T. 2024. Mafia gang find a cash cow in cattle farming to launder money. The Times, 24 July. https://www.thetimes.com/world/europe/article/italian-mafia-gang-cattle-farming-launder-heist-money-dxj2753sh.

Kompaniyets, L., M.A. Evans, and M. Appollonio. 2017. Modeling the relationship between wolf control and cattle depredation. *PLoS ONE*. https://doi.org/10.1371/journal.pone.0187264.

Kreider, R.B., and B. Campbell. 2009. Protein for exercise and recovery. *The Physician and Sportsmedicine* 37 (2): 13–21. https://doi.org/10.3810/psm.2009.06.1705.

Lakhami, N. 2024. America is a factory farming nation': key takeaways from US agriculture census. The Guardian, 15 February. https://www.theguardian.com/environment/2024/feb/15/us-agriculture-census-farming.

Li, M., N. Jia, M. Lenzen, A. Malik, L. Wei, Y. Jin, and D. Raubenheimer. 2022. Global food-miles account for nearly 20% of total food-systems emissions. *Nature Food* 3 (6): 445–453. https://doi.org/10.1038/s43016-022-00531-w.

Li, C., T.R.P. Bishop, F. Imamura, S.J. Sharp, M. Pearce, and S. Brage. 2024. Meat consumption and incident type 2 diabetes: an individual-participant federated meta-analysis of 1·97 million adults with 100000 incident cases from 31 cohorts in 20 countries. *The Lancet Diabetes and Endocrinology* 12 (9): 619–630. https://doi.org/10.1016/S2213-8587(24)00179-7.

Lynch, J., and R. Pierrehumbert. 2019. Climate impacts of cultured meat and beef cattle. *Frontiers in Sustainable Food Systems* 3: Article 5. https://doi.org/10.3389/fsufs.2019.00005.

Madoche, J.L., A. Crosato, E. Mosselman, and S. Maskey. 2024. Effects of urbanization and deforestation on flooding: Case study of Cap-Haïtien City, Haiti. *Journal of Flood Risk Management*, e13020, https://doi.org/10.1111/jfr3.13020.

Mallin, M.A. 2000. Impacts of industrial animal production on rivers and estuaries: Animal-waste lagoons and sprayfields near aquatic environments may significantly degrade water quality and endanger health. *American Scientist* 88 (1): 26–37. https://www.jstor.org/stable/27857960.

Mallin, M.A., V.L. Johnson, S.H. Ensign, and T.A. MacPherson. 2006. Factors contributing to hypoxia in rivers, lakes, and streams. *Limnology and Oceanography* 51 (1, part 2): 690–701.

MARA. 2019. Notice of the Ministry of Natural Resources and the Ministry of Agriculture and Rural Affairs on issues related to the management of land for facility agriculture (自然资源部 农业农村部关于设施农业用地管理有关问题的通知). https://www.gov.cn/xinwen/2019-12/20/content_5462616.htm.
McDermott, J., and A.J. Wyatt. 2017. The role of pulses in sustainable and healthy food systems. *Annals of the New York Academy of Science* 1392: 30–42. https://doi.org/10.1111/nyas.13319.
Micha, R., S.K. Wallace, and D. Mozaffarian. 2010. Red and processed meat consumption and risk of incident coronary heart disease, stroke, and diabetes mellitus: A systematic review and meta-analysis. *Circulation* 121 (21): 2271–2283. https://doi.org/10.1161/CIRCULATIONAHA.109.924977.
Michalak, I., K. Chojnacka, and Korniewicz. 2020. Effect of marine Macroalga Enteromorpha sp. Enriched with Zn(II) and Cu(II) ions on the digestibility, meat quality and carcass characteristics of growing pigs. *Journal of Marine Science and Engineering* 8 (5): 347. https://doi.org/10.3390/jmse8050347.
Miller, V., Reedy, J. Cudhea, F. Zhang, J. Shi, P. Erndt-Marino, J. Coates, J. Micha, R. Webb, P. Mozaffarian, D. Global Dietary Database. Global, regional, and national consumption of animal-source foods between 1990 and 2018: findings from the Global Dietary Database. *Lancet Planet Health*. 2022. Mar;6 (3): e243–e256. https://doi.org/10.1016/S2542-5196(21)00352-1. Erratum in: *Lancet Planet Health*. 2024 Sep 10: S2542-5196(24)00233-X. https://doi.org/10.1016/S2542-5196(24)00233-X.
Mirabelli, M.C., S. Wing, S.W. Marshall, and T.C. Wilcosky. 2006. Asthma symptoms among adolescents who attend public schools that are located near confined swine feeding operations. *Pediatrics* 118 (1): e66–e75. https://doi.org/10.1542/peds.2005-2812.
Mongabay. 2024. The World's Top 10 Biggest Rainforests, Earth.org, 22 June 2024. https://earth.org/worldrainforest-day-worlds-great-rainforests/.
Mullan, S., S.S.C. Nogueira, S. Nogueira-Filho, A. Zanella, N. Rooney, S.D.E. Held, and M. Mendl. 2024. Farming non-typical sentient species: Ethical framework requires passing a high bar. *Journal of Agricultural and Environmental Ethics* 37: 10. https://doi.org/10.1007/s10806-024-09928-y.
NALC. 2024. Animal welfare: An overview. US National Agricultural Law Center. https://nationalaglawcenter.org/overview/animalwelfare/.
NBS. 2024. National data. National Bureau of Statistics of China. https://www.stats.gov.cn/english/.
Nennich, T.D., J.H. Harrison, L.M. VanWieringen, D. Meyer, A.J. Heinrichs, W.P. Weiss, N.R. St Pierre, R.L. Kincaid, D.L. Davidson, and E. Block. 2005. Prediction of manure and nutrient excretion from dairy cattle. *Journal of Dairy Science* 88 (10): 3721–3733. https://doi.org/10.3168/jds.S0022-0302(05)73058-7.

NIH. 2024. Vitamin B6: Fact sheet for consumers. National Institutes of Health Office of Dietary Supplements. https://ods.od.nih.gov/factsheets/VitaminB6-Consumer/.

Nobre, C.A., P.J. Sllers, and J. Shukla. 1991. Amazonian deforestation and regional climate change. *Journal of Climate* 4 (10): 957–988. https://www.jstor.org/stable/26196408.

Nulimaimaiti, M., and H. Huifeng. 2024. China's 'freedom to eat beef' belies farmers' mood of udder chaos as cash cow hits tipping point. South China Morning Post, 12 June. https://www.scmp.com/economy/economic-indicators/article/3266167/chinas-freedom-eat-beef-belies-farmers-mood-udder-chaos-cash-cow-hits-tipping-point.

Nyirabihogo, N. 2024. In DRC, tragic floods demonstrate the lifesaving power of trees. *Global Press Journal*, 20 February. https://globalpressjournal.com/africa/democratic-republic-of-congo/drc-tragic-floods-demonstrate-lifesaving-power-trees/.

Olaniyan, A., and A. Yahaya. 2016. Cows, bandits, and violent conflicts: Understanding cattle rustling in Northern Nigeria. *Africa Spectrum* 51 (3): 93. https://www.jstor.org/stable/44982211.

Papier, K., et al. 2021. Meat consumption and risk of 25 common conditions: Outcome-wide analyses in 475,000 men and women in the UK biobank study. *BMC Medicine* 19: 53. https://doi.org/10.1186/s12916-021-01922-9.

Pooley, S., Barua, M. Beinart, W. Dickman, A. Holmes, G. Lorimer, J. Loveridge, AJ. Macdonald, DW. Marvin, G. Redpath, S. Sillero-Zubiri, C. Zimmermann, A. Milner-Gulland, EJ. An interdisciplinary review of current and future approaches to improving human-predator relations. Conserv Biol. 2017. Jun; 31 (3): 513–523. https://doi.org/10.1111/cobi.12859.

Poore, J., and T. Nemecek. 2018. Reducing food's environmental impacts through producers and consumers. Science—processed by Our World in Data. "Greenhouse gas emissions per 100g protein. https://ourworldindata.org/environmental-impacts-of-food.

Pork. 2024. What are the top 4 states in hog production in the U.S.? *Farm Production Journal*, 25 March 2024. https://www.porkbusiness.com/news/hog-production/what-are-top-4-states-hog-production-us.

Prunier, A., M. Heinonen, and H. Quesnel. 2010. High physiological demands in intensively raised pigs: impact on health and welfare. *Animal: An International Journal of Animal Bioscience* 4 (6): 886–898. https://doi.org/10.1017/S175173111000008X.

Radhakrishnan, S. 2018. A note on wildlife poisoning cases from Kerala, South India. *European Journal of Wildlife Research* 64 (5): 58. https://doi.org/10.1007/s10344-018-1218-6.

RAM. 2024. Global pork meat market forecast report by consuming countries, producing countries, importing countries, exporting countries and company analysis 2024–2032. https://www.researchandmarkets.com/reports/5987485/global-pork-meat-market-forecast-report.

Ritchie, H. 2024. If the world adopted a plant-based diet, we would reduce global agricultural land use from 4 to 1 billion hectares. Published online at OurWorldInData.org. Retrieved from: https://ourworldindata.org/land-use-diets.

Ruckli, A.K., S. Dippel, N. Durec, M. Gebska, J. Guy, J. Helmerichs, C. Leeb, H. Vermeer, and S. Hörtenhuber. 2021. Environmental sustainability assessment of pig farms in selected European countries: Combining LCA and key performance indicators for biodiversity assessment. *Sustainability* 13: 11230. https://doi.org/10.3390/su132011230.

Sasso, A., and G. Latella. 2018. Role of heme iron in the association between red meat consumption and colorectal cancer. Nutrition and Cancer 70 (8): 1173–1183. https://doi.org/10.1080/01635581.2018.1521441.

Savian, M., S. da Penha, and N.M. Holden. 2024. Evaluating environmental, economic, and social aspects of an intensive pig production farm in the south of Brazil: A case study. *The International Journal of Life Cycle Assessment* 28 (11): 1544–1560. https://doi.org/10.1007/s11367-023-02223-4.

Scheidel, A., A. Fernández-Llamazares, A.H. Bara, D. Del Bene, D.M. David-Chavez, E. Fanari, I. Garba, K. Hanaček, J. Liu, J. Martínez-Alier, G. Navas, V. Reyes-García, B. Roy, L. Temper, M.A. Thiri, D. Tran, M. Walter, and K.P. Whyte. 2023. Global impacts of extractive and industrial development projects on Indigenous Peoples' lifeways, lands, and rights. *Science Advantages* 9 (23). https://doi.org/10.1126/sciadv.ade9557.

Schlosberg, D. 2009. *Defining environmental justice: Theories, movements, and nature*. Oxford: Oxford University Press.

Schneider, M. 2017. Wasting the rural: Meat, manure, and the politics of agro-industrialization in contemporary China. *Geoforum* 78 (2017): 89–97. https://doi.org/10.1016/j.geoforum.2015.12.001.

Schultz, A.A., P. Peppaerd, R.E. Gangman, and K.M.C. Malecki. 2019. Residential proximity to concentrated animal feeding operations and allergic and respiratory disease. *Environment International* 130: 104911. https://doi.org/10.1016/j.envint.2019.104911.

Sesnie, S.E., B. Tellman, D. Wrathall, K. McSweeney, E. Nielsen, K. Benessaiah, O. Wang, and L. Rey. 2017. A spatio-temporal analysis of forest loss related to cocaine trafficking in Central America. *Environmental Research Letters* 12 (5). https://doi.org/10.1088/1748-9326/aa6fff.

Shukla, J., C. Nobre, and P. Sellers. 1990. Amazon deforestation and climate change. *Science* 247: 1322–1325.

Smith, B. 2023. China beef update: Domestic production continues to grow as wholesale prices stabilise. Agriculture and Horticulture Development Board. https://ahdb.org.uk/news/china-beef-update-domestic-production-continues-to-grow-as-wholesale-prices-stabilise.

Son, J.-Y., R.L. Muenich, D. Schaffer-Smith, M.L. Miranda, and M.L. Bell. 2021. Distribution of environmental justice metrics for exposure to CAFOs in North Carolina, USA. *Environmental Research* 195: 110862. https://doi.org/10.1016/j.envres.2021.110862.

SPA. 2021. Science panel for the Amazon: Amazon assessment report 2021: Executive summary. https://www.theamazonwewant.org/wp-content/uploads/2022/06/220717-SPA-Executive-Summary-2021-EN.pdf (CC BY-NC-SA 4.0).

Szûcs, E., R. Geers, T. Jezierski, E.N. Sossidou, and D.M. Broom. 2012. Animal welfare in different human cultures, traditions and religious faiths. *Asian-Australasian Journal of Animal Sciences (AJAS)* 25 (11): 1499–1506. https://doi.org/10.5713/ajas.2012.r.02.

Thomas, P. 2023. Smithfield foods CEO defends pork producer's Chinese ownership. The Wall Street Journal, 14 March. https://www.wsj.com/articles/smithfield-foods-ceo-defends-pork-producers-chinese-ownership-a78827da.

Thompson, P.B. 2015. *Form field to fork: Food ethics for everyone*. Oxford: OUP.

Tobvych, V., A. Viazovska, and Y. Kryvoruchko. 2018. Place and people: landscape as a basis of development and preservation of cultural identity. Czasopismo Techniczne 5 (115): s. 65–72. https://doi.org/10.4467/2353737XCT.18.073.8555.

Tondo, L. 2017. How Cosa Nostra's 'cattle mafia' is destroying Sicily's farmers, The Guardian, 14 October. https://www.theguardian.com/world/2017/oct/14/cosa-nostra-cattle-mafia-sicily-farmers-intimidation.

Tudor, K. 2024. Rural crime: Serious organised and international. Who victimises rural communities? National Rural Crime Network. https://nationalruralcrimenetwork.net/rural-crime-report-2024/.

Tuomisto, H.L., et al. 2011. Environmental impacts of cultured meat production. *Environmental Science and Technology* 45 (14). https://doi.org/10.1021/es200130u.

UN FAO. 2006. Livestock's long shadow: Environmental issues and options. Food and Agriculture Organization of the United Nations. https://www.fao.org/docrep/010/a0701e/a0701e.pdf.

UN FAO. 2012. Livestock and landscapes. Food and Agriculture Organization of the United Nations. https://www.fao.org/4/ar591e/ar591e.pdf.

UN GA. 2017. Report of the special rapporteur of the human rights council on the rights of indigenous peoples. A/72/186. https://undocs.org/A/72/186.

USDA. 2024a. Livestock and poultry: World markets and trade. US Department of Agriculture: Foreign Agricultural Service. https://fas.usda.gov/sites/default/files/2024-07/Livestock_poultry.pdf.

USDA. 2024b. Production by commodity. US Department of Agriculture: Foreign Agricultural Service. https://fas.usda.gov/data/production.

Vrdjak, A.F. 2018. Indigenous peoples. *World Heritage and Human Rights, International Journal of Cultural Property* 25 (3): 245–281. https://doi.org/10.1017/S0940739118000218.

Wagner, J.R., F.M. Strong, and C.A. Elvehjem. 1947. Effects of blanching on the retention of ascorbic acid, thiamine, and niacin in vegetables. *Industrial & Engineering Chemistry* 39 (8): 990–993.

Wang, Y., and M.A. Beydoun. 2009. Meat consumption is associated with obesity and central obesity among US adults. *International Journal of Obesity* 33 (6): 621–628. https://doi.org/10.1038/ijo.2009.45.

Wang, B., C. Shen, Y. Cai, D. Liu, and S. Gain. 2022. The purchase willingness of consumers for red meat in China. *Meat Science* 192: 108908. https://doi.org/10.1016/j.meatsci.2022.108908.

WH Group. 2024. Corporate profile. WH Group. https://www.wh-group.com/html/about.php.

White, R. 2022. *Theorising green criminology*. Abingdon: Routledge.

WHO. 2023. Red and processed meat in the context of health and the environment: many shades of red and green. World Health Organisation. https://www.who.int/publications/i/item/9789240074828.

Wiedemann, S., E. McGahan, C. Murphy, M.-J. Yan, B. Henry, G. Thoma, and S. Ledgard. 2015. Environmental impacts and resource use of Australian beef and lamb exported to the USA determined using life cycle assessment. *Journal of Cleaner Production* 94: 67–75. https://doi.org/10.1016/j.jclepro.2015.01.073.

Wilkie, R.M. 2010. *Livestock/deadstock: Working with farm animals from birth to slaughter*. Philadelphia: Temple University Press.

Wing, S., and S. Wolf. 2000. Intensive livestock operations, health, and quality of life among eastern North Carolina residents. *Environmental Health Perspectives* 108 (3): 233–238. https://doi.org/10.1289/ehp.00108233.

Wolfe, R.R. 2017. Branched-chain amino acids and muscle protein synthesis in humans: Myth or reality? *Journal of the International Society of Sports Nutrition* 14: 30. https://doi.org/10.1186/s12970-017-0184-9.

World Bank. 2024. Press release: World Bank supports sustainable development of family farming in Mato Grosso, Brazil, 5 February. World Bank Group. https://www.worldbank.org/en/news/press-release/2024/02/05/world-bank-supports-sustainable-development-of-family-farming-in-mato-grosso-brazil.

Wu, G. 2020. 2020, Important roles of dietary taurine, creatine, carnosine, anserine and 4-hydroxyproline in human nutrition and health. *Amino Acids* 52 (3): 329–360. https://doi.org/10.1007/s00726-020-02823-6.

Wuerther, G. 2010. Assault on nature: CAFOs and biodiversity loss. In *The CAFO reader: The tragedy of industrial animal factories*, ed. by D. Imhoff. Berkeley and Los Angeles: University of California Press.

WWF. 2024. What is human-wildlife conflict and why is it more than just a conservation concern?, Washington, World Wildlife Fund. https://www.worldwildlife.org/stories/what-is-human-wildlife-conflict-and-why-is-it-more-than-just-a-conservation-concern.

Wyatt, T. 2014. A criminological exploration of the industrialisation of pig farming. *Internet Journal of Criminology* 1: 12–28.

Yoshikawa, S. 2023. Agro-pastoral expansion and land use/land cover change dynamics in Mato Grosso, Brazil, Earth, 4 (4): 823. https://doi.org/10.3390/earth4040044.

Zangwill, N. 2021. Our moral duty to eat meat. *Journal of the American Philosophical Association* 7 (3): 295–311. https://doi.org/10.1017/apa.2020.21.

Zeuschner, C.L., B.D. Hokin, K.A. Marsh, A.V. Saunders, M.A. Reid, and M.A. Ramsay. 2013. Vitamin B12 and vegetarian diets. *Medical Journal of Australia* 199 (54): S27–S32. https://doi.org/10.5694/mja11.11509.

Zhang, S., X. Wu, D. Han, Y. Hou, J. Tan, S.W. Kim, D. Li, Y. Win, and J. Wang. 2021. Pork production systems in China: A review of their development, challenges and prospects in green production. *Frontiers of Agricultural Science and Engineering* 8 (1): 15–24. https://www.doi.org/https://doi.org/10.15302/J-FASE-2020377.

CHAPTER 3

Fruit and Vegetables

Abstract This chapter investigates the hypothetical cessation of global tomato and potato trade, framing the discussion through the lenses of environmental, ecological, species, and economic justice. Tomatoes and potatoes, cornerstones of diets and agricultural economies worldwide, are emblematic of the interconnected nature of food systems. The chapter explores their historical origins, genetic diversity, and the resource-intensive practices behind their cultivation. Industrial farming methods, including monoculture and irrigation, are scrutinised for their environmental impacts, such as soil degradation, water scarcity, and biodiversity loss. The discussion also addresses ethical concerns surrounding genetic modification and the exploitation of migrant labour, particularly in Europe, China, and the United States. Economic implications, including disruptions to global trade and smallholder livelihoods, are analysed alongside potential shifts towards localised food sovereignty and sustainable farming methods. Emerging technologies, such as controlled-environment agriculture and genetic innovations, are presented as viable solutions to mitigate the environmental and ethical challenges posed by industrial-scale production. Ultimately, the chapter underscores the need for systemic reform to create equitable, sustainable, and resilient food systems.

© The Author(s), under exclusive license to Springer Nature Switzerland AG 2025

S. Sneddon, *Food, Environmental Degradation and Injustices*, Palgrave Studies in Green Criminology,
https://doi.org/10.1007/978-3-031-87900-5_3

Keywords Monoculture · Genetic modification · Biodiversity loss · Controlled-environment agriculture · Migrant labour exploitation

Introduction

What would happen if two of the world's most ubiquitous crops—tomatoes (*Solanum lycopersicum*) and potatoes (*Solanum tuberosum*)—suddenly ceased to be traded? This chapter looks at the effects of such a scenario through the lenses of environmental, ecological, species, and economic justice. These crops, essential not only to diets worldwide but also to agricultural economies and cultural identities, neatly demonstrate the interconnectedness of food systems.

Tomatoes and potatoes occupy a unique place in the agricultural world. They are among the few crops that are cultivated extensively on both industrial and small scales, from sprawling commercial farms to tiny home gardens. Globally, there are an estimated 10,000 cultivars of tomatoes and around 5000 varieties of potatoes, highlighting their remarkable genetic diversity. These range from the minuscule "Tomberry," weighing just 1–2 g per fruit (Littletom 2016), to the record-breaking "9.06 Brown," which produced a 5.2-kilogram tomato (GWR 2024). Potatoes, too, span a wide spectrum of colours, shapes, and uses, making them staples in cuisines from Asia to Europe.

Both crops belong to the Solanaceae family, sharing lineage with other staple foodstuffs including aubergines (eggplants) (*Solanum melongena*) and bell and chilli peppers (*Capsicum annuum*), decorative plants like the Chilean potato vine (*Solanum crispum*), psychoactive plants such as Mandrake (*Mandragora officinarum*), as well as toxic plants like deadly nightshade (*Atropa belladonna*). Despite their importance today, they have relatively recent histories in global agriculture. Originating in Central and South America, tomatoes and potatoes were domesticated by Indigenous peoples and then spread worldwide following the Spanish conquest of the Americas in the sixteenth century. Initially met with suspicion in Europe due to their association with poisonous plants, these crops eventually found acceptance and now underpin diets across continents.

In modern times, the commercial significance of tomatoes and potatoes cannot be overstated. In 2023, the global tomato market was valued at $195 billion (Straits Research 2023), while the potato market was worth

$155 billion (Indexbox 2024). These figures only capture their economic footprint in formal trade; countless individuals also cultivate them for personal use, underscoring their accessibility and versatility. However, the potential cessation of their trade poses important questions: How would the loss of these crops impact food systems, biodiversity, labour practices, and global economies?

From an environmental justice perspective, tomatoes and potatoes illustrate the resource-intensive nature of industrial agriculture. Tomatoes are among the most water-demanding crops, requiring nearly 2000 l of water per 1000 calories produced. Potatoes are less water-intensive but still present challenges in regions where irrigation strains local water supplies. Monoculture farming practices, prevalent in the production of both crops, exacerbate soil degradation and biodiversity loss while increasing reliance on synthetic fertilisers and pesticides. In arid regions, the environmental cost of these practices is even higher, highlighting the inequities in how natural resources are used and distributed.

The ecological justice implications are equally significant. The heavy use of pesticides in tomato and potato farming has far-reaching consequences for ecosystems, from contaminating water supplies to harming pollinators like bees. Furthermore, monoculture farming disrupts local biodiversity, creating imbalances that affect entire ecosystems. The potential for genetic contamination between genetically modified (GM) varieties and wild relatives also raises concerns about the long-term health of natural plant populations. If these crops were to vanish from trade, one might expect some relief for strained ecosystems, but the adjustment period could introduce new challenges, including shifts to alternative crops that might not be more sustainable.

Species justice considerations also come into play. Tomatoes and potatoes are emblematic of the tension between modern agricultural practices and biodiversity preservation. Over centuries, traditional farming practices nurtured a wide array of varieties adapted to specific climates and cultural needs. However, the rise of commercial agriculture has narrowed this diversity in favour of high-yield monocultures. This homogenisation not only threatens crop resilience but also diminishes the cultural heritage embodied in "heritage" and indigenous varieties. The sudden cessation of trade in these crops might inadvertently provide opportunities to revive and prioritise localised, diverse farming systems that better align with species justice principles.

Economically, the ramifications of losing these crops from global trade would be profound. Tomatoes and potatoes are essential income sources for millions of farmers worldwide, from smallholders in developing nations to large-scale producers in industrialised countries. Their production supports extensive supply chains, encompassing seed and chemical manufacturers, transportation networks, and retail operations. The loss of these crops would disrupt livelihoods, increase food prices, and create a void in agricultural markets that would take years to fill. Furthermore, countries that rely heavily on their export, such as China for potatoes or Italy for processed tomato products, would face significant economic upheaval.

The potential cessation of tomato and potato trade also has implications for food sovereignty—the right of communities to define their own food systems. These crops are among the few that are relatively easy to grow on a small scale, making them accessible even to amateur gardeners. A shift towards localised cultivation could mitigate some of the economic and food security impacts of losing these crops from formal trade channels. However, this transition would require substantial investments in education, infrastructure, and support for small-scale farmers to ensure equitable access to resources and markets.

This chapter explores these themes and examines how the hypothetical loss of tomatoes and potatoes from global trade would affect environmental, ecological, species, and economic justice. This exercise is not merely speculative; it serves as a lens through which to evaluate the vulnerabilities and inequities inherent in our food systems. By understanding the far-reaching consequences of losing even two crops, we can better appreciate the need for resilient, sustainable, and just agricultural practices.

Environmental Justice

While the commercial growing of tomatoes and potatoes does not raise the same ethical issues as those covered in other chapters (particularly Chapters 2 and 4), there are nonetheless associated ethical risks. These fall into two main categories, the ethical issues around the use of genetic modification and the ethical issues around the use of migrant labour.

Genetic Modification

For both tomatoes and potatoes (and many other foodstuffs), there have always been ethical debates around the use of genetic modification. The very fact that there are many thousands of types of tomato and potato is evidence that the original, wild stains of both plants have been selectively crossbred over the centuries to achieve certain characteristics. These may relate to colour, taste, size, speed of ripening, ability to withstand certain climatic conditions, or pest and disease resistance. Historically, this has been done by cross-fertilising two types of the same crop with the traits that are desired and slowly developing a new type (Hill 2013). This raises no particular ethical issues and has been common practice globally for centuries.

The two types of modification which have raised ethical issues are "cis- or transgenic gene modification (i.e., insertion of (foreign) gene material) and genome editing (i.e., insertion, deletion, modification or replacing of gene material)" (Bearth et al. 2022, p. 1118). Early in the development of genetic modification, the Nuffield Council on Bioethics identified that there were three main ethical concerns about the use of generic modification in foods:

- Possible harm to human health;
- Possible damage to the environment; and
- The "unnatural" status of the technology (Kennedy 1999, p. 2).

In addition to these early concerns, as the GM market has matured, it has become apparent that there is a distinctly unequal relationship between small farmers and the multinational seed companies. Research by Clapp (2015) "shows that an increasingly small number of powerful firms have come to play an enormous role in shaping the dominant agricultural model adopted around the world. This influence has resulted in a further narrowing of crop genetic diversity and the increased use of agro-chemicals" (Clapp 2015, p. 28).

Tomatoes have the distinction of being the first genetically modified foodstuff to be given a licence for human consumption. In 1994, the California-based biotech company Calgene (bought by Monsanto in 1997 and now part of Bayer) gained US Food and Drug Administration approval for its genetically modified tomato known as CGN-89564-2, or Flavr Savr™ (Kramer and Redenbaugh 1994; Martineau 2001). In 2015, a type of potato was approved by the USDA which has been genetically engineered to minimise the risk of blemishes and browning once cut (Waltz 2015a). This was unusual at the time, because no "foreign" genetic material was introduced; rather, a gene was removed and then reinserted (Waltz 2015b).

One of the best ways to ensure that crops will flourish in specific areas is to save and replant seeds. This will ensure that "heritage" varieties (older varieties, which may not be as productive as modern commercial hybrid varieties) are not lost and preserve the widest possible genetic diversity in particular types of crop. However, with the increasing centralisation of power by the multinational seed companies, the ability of smaller farms to save seeds and preserve this biodiversity is declining.

It is easy to see how this corporate dominance could threaten local food systems and cultural heritage, infringing upon the environmental justice principle of recognition. Howard (2009) identifies the inherent conflict at play here, saying that "renewable agricultural practices of seed saving and replanting are nevertheless at direct odds with increasing profits in the global seed industry" (Howard 2009, p. 1281). This is amplified with GM crops, which are patented and are generally designed not to reproduce naturally.

The biodiversity concerns around GM crops and biotechnology in general are addressed in the Cartagena Protocol on Biosafety to the UN Convention on Biological Diversity, which was created in 2000. The Protocol covers the transboundary movement of "Living Modified Organisms" or LMOs, which is defines as "any living organism that possesses a novel combination of genetic material obtained through the use of modern biotechnology" (Article 3(g)). The Protocol requires that "the development, handling, transport, use, transfer and release of any living modified organisms are undertaken in a manner that prevents or reduces the risks to biological diversity, taking also into account risks to human health" (Article 2(2)) but further empowers them "to take action that is more protective" (Article 2(4)) provided such action does not contravene international law. Interestingly, despite being in existence for nearly a quarter of a century, Odong (2024) argues that there are still inherent and unresolved conflicts between the trade restrictive practices of the Protocol and rules of the General Agreement on Tariffs and Trade (GATT)/World Trade Organization (WTO), and argues that the Protocol must be given "legal teeth" as it "must fulfil its potential of shielding the environment from the adverse effect of transboundary movement of LMOs" (Odong 2024, p. 15).

For all crops, including tomatoes and potatoes, the proponents of GM argue that it will allow disease-resistant strains to be developed which will also tolerate different climactic conditions—essential given the rise in global temperatures. Indeed, Abushal et al. (2021) strike a resoundingly positive note about the use of all forms of genetic modification, concluding that "even though there might be bioethical concerns regarding agricultural biotechnology and GMOs, there is sufficient empirical evidence to state that this human endeavor has the capacity and capability to promote the wellbeing of society" (Abushal et al. 2021, p. 7). The FAO (2024) is not as positive, concluding that the developments in GM crops and other forms of biotechnology "have not sufficiently benefited smallholder farmers and producers and consumers" (FAO 2024) and that changes to regulations are needed, since "to produce food in a sustainable way for an additional 2 billion people by 2050, a business-as-usual approach will not be sufficient" (FAO 2024).

From an environmental justice perspective, the introduction of GM crops without the informed consent of affected communities violates the principles of participation and recognition (White 2014). It is also counter-productive and short-sighted, as disregards the knowledge and values of Indigenous populations, and exclusionary to small-scale farmers who may have reservations about genetic modification due to cultural or ecological reasons.

Migrant Workers

For tomatoes, there have been strong ethical concerns raised about the treatment of migrant workers, especially in Italy and Spain. The cultivation of potatoes and tomatoes is a labour-intensive and often poorly paid process, and often the labour demands cannot be met locally. In Spain's Almería region, for example, "each hectare requires an average of 560 working days per year, the equivalent of 2.3 full-time workers" (Aznar-Sánchez et al. 2014, p. 118).

This means that the use, and exploitation of vulnerable migrant workers, is a significant risk. In Italy, caporale ("corporals") are gangmasters operating at the behest of farmers, and largely employ migrant workers—this process is known as "caporalato" (Perrotta 2019). With workers outnumbering work, the opportunities for unscrupulous caporale to engage in exploitative practices are huge (Bagnardi et al. 2022). A 2012 report by Amnesty International found that migrant workers were being regularly exploited, with "wages considerably below the minimum agreed by unions and employers, arbitrary reductions of wages, delays or non-payment of wages and very long hours of work" (Amnesty International 2012, p. 37). In the Almería region of southern Spain sits the Campo del Dalías "which has one of the highest concentrations of greenhouses in the world" (NASA 2022). In addition, Aznar-Sánchez et al. (2014) point out that Almería has "Spain's second highest percentage of migrants" (2014, p. 112). Initially, this led to a high proportion of undocumented migrants, who had little or no protection and, although Aznar-Sánchez et al. (2014) say that laws in Spain have been tightened to reduce the flow of undocumented migrants, more recent global events have meant that the numbers have increased again with net migration (documented and undocumented) in 2022 alone "surpassing that of the entire previous decade" (Reuters 2024). Indeed, there are so many undocumented migrants in the country that the government announced in November 2024 that it would issue work or residence permits to around 300,000 people a year, for the next three years (Reuters 2024). As a positive, this move could provide many of the currently exploited undocumented migrant workers with higher levels of protection, and reduce the impact of organised immigration crime, although it will also increase the costs associated with employing them. Italy and Spain are not the only examples of migrant farmworkers being exploited in the tomato and potato industries. In the United States, Gold et al. found that 7% of farm workers in 2019–2020 were international migrants (2022, p. 7), although Human Rights Watch has previously documented that "the agricultural industry relies heavily on unauthorized immigrants, who make up about 50% of the workforce, if not more" (2012, p. 6). If there is a disparity in these figures, it is likely because undocumented/unauthorised migrants would be less likely to engage with a formal survey by a US government department (the Department of Agriculture).

China too has faced accusations of violations of the rights of workers in the tomato industry. In 2021, the United States banned all imports from Xinjiang Province in China by passing the Uyghur Forced Labor Prevention Act (Public Law No: 117-78) which changes US policy by creating a rebuttable presumption that all products (which include tomatoes) produced in Xinjiang are produced using the forced labour of Uighur and other minority groups. The use of forced labour in the region is also the subject of a UN Office of the High Commissioner for Human Rights report (OCHCR 2022) which concludes that:

> The extent of arbitrary and discriminatory detention of members of Uyghur and other predominantly Muslim groups, pursuant to law and policy, in context of restrictions and deprivation more generally of fundamental rights enjoyed individually and collectively, may constitute international crimes, in particular crimes against humanity. (OCHCR 2022, para 148)

In 2024, a BBC Eye investigation found that many types of tomato purée sold in the UK and Germany and labelled at "Italian-produced" (or similar) were actually imported as tomato paste into Italy from the Xinjiang region (BBC Eye 2024). In the EU, Directive 2024/1760 on Corporate Sustainability and Due Diligence was passed in the summer of 2024, and requires Member States to pass legislation ensuring that all companies over a particular size carry out due diligence checks "regarding actual and potential human rights adverse impacts and environmental adverse impacts, with respect to their own operations, the operations of their subsidiaries, and the operations carried out by their business partners in the chains of activities of those companies" (Article 1(1)(a)).

Human Health

There are no specific health issues associated with most consumption of tomatoes or potatoes (barring allergic reactions). Tomatoes are a good sources of retinol (Vitamin A), which helps with vision, immune systems, cell growth, and heart and lung development (NHS 2024), ascorbic acid (Vitamin C) which prevents scurvy and helps to maintain healthy skin, bones, and cartilage (NHS 2024) and lycopene, an antioxidant which helps to prevent heart attacks and fatty liver disease (Stice et al (2018). Potatoes are also high in ascorbic acid and are also a good source of fibre and carbohydrates. Improperly stored potatoes can develop green colouration, meaning that they are high in solanine (Friedman and McDonald 1997). Solanine (which is also present in Deadly Nightshade) is a glycoalkaloid toxin which the plant uses as a defence but which, when consumed by humans, can result in health conditions ranging from vomiting and diarrhoea to death, depending on the dosage (AGES 2024).

Although the proper consumption of tomatoes and potatoes poses very little by way of health risks, their production can involve significant use of pesticides. Pesticide exposure can lead to acute poisoning, with symptoms ranging from headaches and nausea to respiratory distress and, in severe cases, death (Mostafalou and Abdollahi 2017). Chronic pesticide exposure has been linked to long-term health issues such as cancer, neurological disorders, and reproductive problems (Kim et al. 2017). In many developing countries, the regulations may be lacking, or less strict, and the enforcement of any regulations may be weaker. Farmworkers often lack access to protective equipment and training on safe handling practices (Kesavachandran et al. 2009). Additionally, communities living near agricultural areas may be exposed to pesticide drift, contaminating air, water, and soil (Aktar et al. 2009). Jørs et al. (2006) looked at pesticide use by farmers in Bolivia, and "documented a frequent use of the most toxic pesticides among farmers who have had almost no instructions in how to use pesticides and protect themselves against the dangers of intoxication, reflected in the hazardous practices used when handling pesticides."

In conclusion, while the commercial cultivation of tomatoes and potatoes may not present the same ethical challenges as some other agricultural sectors, significant ethical risks persist, particularly concerning genetic modification and the exploitation of migrant labour. The use of advanced genetic modification techniques raises concerns about potential harm to human health, environmental damage, and the erosion of biodiversity due to corporate dominance in the seed industry. This corporate control threatens local food systems and disregards the principles of environmental justice by marginalising small-scale farmers and Indigenous communities. Simultaneously, the reliance on vulnerable migrant workers in countries like Italy, Spain, the United States, and China has led to widespread exploitation, with inadequate labour protections and poor working conditions. Migrant worker exploitation is, of course, much more widespread than these two industries; rather, they have been identified as "good opportunities" for OCGs and others to exploit. To a large extent, these negative impacts are offset by the overall beneficial impact of tomato and potato consumption.

Ecological and Species Justice

The water requirements of tomatoes make them the third-highest foodstuff in terms of water use behind farmed prawns and farmed fish. It is the most water-intensive land-based crop. The carbon footprint of tomatoes depends upon where, when, and how it is grown. Heated greenhouses will likely use more energy but will have a longer growing season than outdoor-grown crops in many parts of the world. It is also possible to successfully grow tomatoes in soil-free indoor environments under full spectrum LED lighting, which means that in the future the demand for land, pesticides, herbicides, and fertiliser could be significantly reduced.

In terms of the carbon emissions, land use, and water use per 1000 kCal, the worst-performing vegetables are tomatoes (11 kg/ 4.21 m^2/1946 l) and potatoes (0.63 kg/1.2 m^2/81 l) (Ritchie et al. 2024). It is immediately apparent that despite being the worst-performing vegetables, these figures are far lower than the equivalent figures for beef and pork, as explored in Chapter 2. Potatoes appear to be an all-round winner as far as foodstuffs are concerned. They have one of the lowest carbon footprints, water usage, and land usage of any crop and, in addition, Tunio et al. (2020) strongly argue that potatoes are suitable for aeroponics (indoor, soil-free growing), with similar benefits to tomatoes.

Space, Place, and Waste

In the previous chapter, we saw how the expansion of farming of cattle and pigs was leading to significant habitat destruction. The primary focus of much of global attention is, with good reason, the deforestation of the Amazon rainforest and, to a much lesser extent, the equally dramatic deforestation in the Congo Basin rainforest. Habitat destruction for tomato and potato farming is considerably less headline-worthy, but equally significant. Foley et al. (2011) made the point that "approximately a billion people are chronically malnourished while our agricultural systems are concurrently degrading land, water, biodiversity and climate on a global scale" (2011, p. 1). Over a decade later, nothing significant has changed, and agriculture is still the largest contributor to habitat loss, and the "second greatest global threat to preserving biodiversity" (Aznar-Sánchez et al. 2014, p. 665). As we saw above, the Almería region of Spain has the largest concentration of greenhouses in the world, and while these have "resulted in very significant social and economic benefits for the Almería Province" (Castro et al. 2019, p. 3), they have also had "important negative impacts on native biodiversity and natural resources" (Castro et al. 2019, p. 3). FAOSTAT, the statistics body of the UN Food and Agriculture Organization, calculated that in 2022 tomato growing covered 5 million hectares (50,000 km^2) of agricultural land (FAOSTAT 2022), and potato growing covered 18 million hectares 180,000 km^2) (Dongyu 2022). While 230,000 km^2 may not seem like an enormous space, this is approximately the same land area as the whole of the United Kingdom. In 2008, Hajjar et al. concluded that "the conservation of this [genetic] diversity stabilizes ecosystem functioning in the long-run" and the evidence for this has only grown stronger in the intervening years (see, inter alia, Brooker et al. 2021; Evans et al. 2022). Soil erosion also hampers the ability of soil to store, filter, degrade, or inactivate some of the functions of pesticides (Burauel and Baßmann 2005). Lal (2001) notes that irrigation and the use of fertilisers will mask the inherent degradation of the soil caused by agriculture, and further that soil degradation leads directly to "reduction in crop yields and decline in their nutritional values (protein content, micronutrients etc.)" (Lal 2009, p. 46). Indirectly, soil degradation leads to "reduction in use efficiency of inputs (e.g., fertilizer, irrigation water) and additional land area required to compensate the loss of production" (Lal 2009, p. 46).

Monoculture, which is the preferred technique for commercial potato and tomato production favouring high-yield varieties, has a host of environmental problems, from reduced resistance to the invasion of introduced species (Tscharntke et al. 2005) to depleted soil quality (Bogužas et al. 2022). These in turn lead to a perceived need for increased use of pesticides and nitrogen-based fertilisers, bringing with them a further deterioration of the environment. Equally, the fact that tomatoes grow better in warmer climates means that it is likely irrigation will be used to satisfy the high demand for water. Hoekstra and Chapagain (2007) used Allan's (1993) concept of "virtual water" which includes not only the water which is used to hydrate the crop, but also the water which is embedded within the production and supply of the crop. Extrapolating figures from Hoekstra and Chapagain's work (2007, p. 41), tomatoes require 185 l of "virtual water" ("LVW") to produce 1 kg of fruit, and 1 kg of potatoes require 250 LVW. Combined with the average calorific content for both fruit, this gives a demand of 1027 LVW water per 1000 kCal for tomatoes and 266 LVW per 1000 kCal for potatoes. What is interesting is the lack of specific correlation with the figures from

Ritchie et al. (2024) about physical water use, although tomatoes still require more water and LVW than potatoes.

In several arid and semi-arid areas across the globe, the irrigation demands of farming have caused significant environmental issues. Thirty years ago, Postel et al. (1996) calculated that "Fresh water is now scarce in many regions of the world, resulting in severe ecological degradation, limits on agricultural and industrial production, threats to human health, and increased potential for international conflict" (p. 786), and in 2021, the Office of the Director of National Intelligence confirmed that "the next two decades, declining water security is likely to exacerbate existing social grievances and divisions, potentially triggering or worsening conflict between societal groups and industry sectors and increasing the risk of political instability" (ODNI 2021).

In addition to increased irrigation, monoculture crops tend to see greater use of pesticides, whereas multicropping can rely on natural systems to control some pests and reduce the need for chemicals. Guinet et al. (2023) looked at "temporal crop diversification" which covers alternating species of crops, using cover crops and a more "patchwork" landscape of crops, and concluded that this method could reduce pesticide use. Currently, tomatoes and potatoes are predominantly grown as monocrops, and this requires an increased use of pesticides. The most commonly use class of pesticide globally is neonicotinoids, which are highly effective at controlling a wide range of insects. Unfortunately, neonicotinoids are also highly effective at killing bees and other pollinating insects (see, for example, Goulson 2013; Ollerton et al. 2014; Aguiar et al. 2023), and this has led to global concerns about their use. Sánchez-Bayo and Wychuys (2019) have calculated that 40% of the world's insect species may become extinct by the end of the 2020s, and that the two main drivers are habitat loss and the use of synthetic pesticides and insecticides—both of these are fundamental issues in the farming of tomatoes, potatoes, and other vegetable crops. The use of three categories of neonicotinoid has been banned in the European Union since 2018[1] and attempts by some EU Member States to derogate from this ban were ruled to be illegal by the Court of Justice of the EU in 2023 (C-162-21). The secondary risk from overuse of insecticides is that

[1] The bans are on imidacloprid (http://data.europa.eu/eli/reg_impl/2018/783/oj), clothianidin (http://data.europa.eu/eli/reg_impl/2018/784/oj), and thiamethoxam (http://data.europa.eu/eli/reg_impl/2018/785/oj).

species may develop a resistance to specific chemicals, in the same way as antibiotic use causes resistance as discussed in Chapter 2. Bade and Bhamare (2024) revealed that cotton bollworm (Helicoverpa armigera), which is a significant chickpea pest, had developed a sevenfold increase in resistance to an insecticide used to control it. Similarly, other studies have found increased insecticidal resistance in pulse beetles (Gupta et al. 2023), whitefly (Prasad et al. 2022), and aphids (Song et al. 2024).

The use of genetic modification also poses both potential and real risks in terms of ecological justice. Just like pesticides and insecticides, which protect plant species by eliminating species which prey upon them, tomatoes and potatoes have been genetically modified to contain insecticidal toxins, among other substances. Pizetta et al. (2022) looked at GM tomatoes which had had genetic material inserted to make then tolerant to predation by whitefly (Bemisia tabaci), an insect which causes considerable damage by sucking the sap from tomato plants. From the perspective of the tomato grower, this modification is useful, as it reduces crop losses. However, whitefly themselves are food for a range of other species, such as parasitic wasps, and mites (RHS 2024), and so controlling their numbers will have a knock-on impact elsewhere.

The longer-term ecological risk of GM crops is that there will be unintended and uncontrolled crossbreeding between the GM variant and naturally occurring variants. Keese (2008) say that the possibility of horizontal gene transfer (HGT) between plants is low, but "Even when the frequency of HGT may be high, a close-to-zero probability of any of the other steps will reduce the overall likelihood that harm is realized to also near zero" (2008, p. 132). Keese concludes that the risk to human health from HGT from GM plants is negligible. The Royal Society, however, which has previously reviewed and debunked research suggesting GMOs were harmful to health in laboratory test animals, states unequivocally that "GM crops may cross breed with closely related plants. This includes non–GM varieties of the same crop and wild relatives of the crop" (2024). The Royal Society too, while differing in their view on the likelihood of HGT from GM crops, agrees that the risk to health and the environment is extremely low. Unlike the ban on nicotinoids, where the CJEU ruled that Member States had legal ability to derogate, the approach taken by the EU in relation to GM crops and other GMOs is that Member States are free to decide if their territory is excluded from EFSA authorisations, or from cultivation of authorised GMOs (EFSA 2024).

Potato farming and tomato farming present significant ecological justice challenges, including habitat destruction, soil degradation, water resource depletion, pesticide impacts, loss of biodiversity, and potential ecological risks from genetic modification. Often, short-term agricultural and financial gains are prioritised over the long-term health and integrity of ecosystems. In order to achieve ecological justice, agricultural systems need to ensure that the ecosystems in which the processes take place are not compromised, and the current use of fertilisers, pesticides, irrigation, and monocropping is clearly not maintaining that balance. From a strictly ecological and species justice perspective therefore, the ending of commercial farming of these two crops would be a net benefit. As ever, however, the economic impacts of such a ban are likely to be significant and will be explored in the following section.

Economics

FAOSTAT report that despite technically being a fruit, tomatoes were the most produced vegetable globally in 2022, with 186 million tons produced (FAOSTAT 2023, p. 9), which is a slight dip from 2021. Potatoes were "the most produced commodity" in the tubers category, with 375 million tons produced in 2022 (FAOSTAT, p. 8). Both crops are fundamental parts of diets and cuisine worldwide, and linked to this are issues around income for farmers, especially slam-scale farmers, workers, and global trade. The World Bank identifies that "agricultural development is one of the most powerful tools to end extreme poverty, boost shared prosperity and feed a projected 10 billion people by 2050" (World Bank 2024) and Abraham and Pingali (2020) note that while UN Sustainable Development Goal 2 (Zero Hunger) is explicitly linked to smallholder farming, "eight other goals related to ending poverty, gender discrimination, inequality and environmental degradation, tackling climate change, and promoting and ensuring healthy lives have small farm development and growth central to their success" (2020, p. 174). Tomatoes and potatoes are significant sources of income for farmers in China, India, the United States, and elsewhere, and the loss of these crops would require diversification into other crops, which will have a significant short- and medium-term impact on income levels.

In addition to the financial impact on farmers, both large and small, tomato and potato production involves a significant upstream and downstream supply chain. This will include the companies producing the seeds, chemicals, and agricultural machinery used to grow the crops, to the workers (both legal and illegal), to the shops and restaurants which would sell the products. Intervening in this supply chain "should be designed in order not to reduce economic efficiencies or produce unforeseen consequences" (Truelove et al. 2023, p. 108).

Having demonstrated in Chapter 2 that the economic impact of ceasing beef and pork production would be felt more significantly in scale by the animal-industrial complex, but more significantly in proportion by smaller farms, we will now look at the ***distribution*** of the economic impacts of ceasing tomato and potato farming. To do that, the largest producer regions in the largest producer countries will be identified, and the income differential between tomato and potato farmers and other state citizens will be assessed.

The largest producer of tomatoes in the world is China, which produced more than a third of the total crop (FAOSTAT 2023). As we have seen, above, the region of China which is most suitable for tomato growing is the Xinjiang Uygur Autonomous Region (Xinjiang), in the far northeast of the country, bordering on nine countries and the Tibet Autonomous Region. The GDP per capita in 2023 was CNY73,774 (approximately US$10,150), which puts the predominantly Uyghur Province at 16th highest (out of 31), at 36% of the level of the richest province, Beijing (NBS 2024). This is not the full picture, however, as there is significant divergence between the wealthiest in the Region and the poorest. The city of Karamay, with just under half a million predominantly Han Chinese inhabitants, is the third richest city per capita in the whole country, behind the Special Administrative Regions of Hong Kong and Macau. At the other end of the scale are the Uygur detainees in "re-education" camps who are alleged to be subjected to forced labour (OCHCR 2022). Exports of products produced in Xinjiang are already banned in the United States, as we saw earlier, and any using forced labour are likely to fall under the both the UK's Procurement Act 2023 (which comes into force in 2025) and the EU Directive on Corporate Sustainability and Due Diligence, which comes into force in 2026. Any impact from stopping tomato production would have marginal impact on the poorest in the region, since it is alleged they are not paid anyway and, since the wealthiest derive a substantial element of their income from the significant coal, oil, and gas deposits in the region, it is unlikely to cause negative direct impacts on them either.

The largest producer of potatoes in the world is also China, which produces around one third of the world's potato crop (FAOSTAT 2024). In 2016, China announced that rather than focusing of the production of potatoes for starch, the government was promoting the idea of potato as a staple food (Wang et al. 2023, p. 556). The climate and geography of China mean that potatoes can be grown in almost all regions, but Wang et al. (2018) use Chinese Ministry of Agriculture data to show that the north of the country (which produces a summer crop) and the southwest (which produces both a summer and winter crop) produce nearly 90% of the total. Without specific grow sites, it is harder to draw the economic parallels, but Zheng et al. (2024) show that potato farmers across China "have an annual per capita income of 5000 to 10,000 yuan" (US$650–1300), which is between 5 and 10% of the national GDP per capita, at US$12,600. Unlike with tomatoes, there is no evidence of significant labour exploitation connected to potato farming. Country-wide, China exports only a fraction of its production, around US$200 m in 2023 (TrendEconomy 2024) compared to the Netherlands which exports a fifth of the world's potatoes, worth US$1.2bn. An end to this trade would have a much more significant impact on the international trade of the Netherlands than of China, but even then, the impact would be negligible—the Netherlands exported €870bn (US$920bn) of goods in 2023 (Statistics Netherlands 2024) so potato exports are less than one-tenth of one per cent of the total.

Unlike beef and pork (Chapter 2) or salmon, tilapia, and prawns (Chapter 4), a cessation in industrial production of tomatoes and potatoes would not necessarily lock consumers in to quite such a stark binary choice between decreased consumption and increased expenditure on organic produce, as it is possible to grow both tomatoes and potatoes in a domestic setting. In most circumstances, however, it is not feasible to grow the same volume of either crop as would previously have been purchased. Allied to this is the concept of "food sovereignty" which gaining traction across the world, although it is a contested term (see Patel 2009). The Nyéléni Declaration from 2007 summarised food sovereignty as "the right of peoples to healthy and culturally appropriate food produced through ecologically sound and sustainable methods, and their right to define their own food and agriculture systems." While it is acknowledged that "home grown" is not necessarily the same as "food sovereignty," it is a small but crucial step in consumers gaining more power over the production of their food and may go some way to mitigate changes in the supply chain.

Overall, we can see the cessation of industrial tomato and potato farming worldwide would trigger substantial economic repercussions. These are staple crops which underpin livelihoods, from small-scale farmers striving to improve incomes and meet development goals, to large agribusinesses and extended supply chains supporting seed production, chemical inputs, transportation, and retail. As demonstrated, the impacts for potatoes would be more unevenly distributed than those for tomatoes. However, this is a global agribusiness, linked to international trade, geographical disparities, pre-existing socio-economic inequalities, and the presence of forced labour or contested political conditions, such as those alleged in China's Xinjiang region.

While ceasing production could encourage diversification, improved sustainability, and even enhance certain forms of food sovereignty, it would also threaten food security, spur higher prices for alternative staples, and erode farmer incomes. From an economic standpoint therefore, this would not be a viable option without substantial prior effort being put into the transition to future developments in production methods. The final part of the chapter will raise some of these developments and discuss the potential that they may have for mitigating the end of the commercial production of tomatoes and potatoes.

Future

There are several factors which are likely to play a role in the future of commercial tomato and potato farming, and this section will discuss the two which are likely to be most significant. Firstly, climate change, bringing with it unpredictable weather patterns, increasingly powerful storm systems and changes in pollinator populations. As part of this, the continuing development of biotechnology and the use and creation of new genetic modifications are likely to play a role, although this may be tempered somewhat by any continuing consumer scepticism and fear. The second factor is the likely improvements in farming practices, especially hydroponics and indoor growing, which will bring with it reduced demand for outdoor farming land.

Climate Change and Genetic Modification

In a Special Report on the impact of climate change on land, the Intergovernmental Panel on Climate Change suggested that food productivity is likely to be "negatively affected by increased pests and diseases, as well as changing distributions of pollinators under climate change" (IPCC 2019, p 523). Both tomatoes and potatoes are known to suffer from heat and water stress, meaning the hotter the temperature and the lower the available water, the poorer the crop and the more susceptible the plants are to pests and diseases. In tomatoes, Bita and Gerats (2013) have demonstrated that increases in temperature can induce heat stress, and reduce fruit set (the point where a pollinated flower develops into a fruit), and compromise fruit quality. Naresh Kumar et al. (2015) have calculated that in parts of India. Climate change is likely to decrease the total crop but interestingly, Haverkort and Struik (2015) conclude that in "most countries, yields of potato might increase by climate change, provided water supply remains adequate" (p. 76). What this does mean is that considerable effort is likely being expended in order to breed and develop strains of both crops which will prove more resilient to the impact of climate change. For example, the EU-funded TomGEM project, looked at genetic markers which would make tomatoes more resilient, and "main outcomes of the project are the superior genotypes in terms of tolerance to heat stress and yield stability, but also new genes and new markers that will help the tomato breeders to create new tomato varieties and cultivars that are better suited to high temperature conditions" (TomGEM 2021, p. 4). For potatoes, Meacham-Hensold et al. (2024) genetically altered Desiree potatoes to create a variety which was larger, did not lose any nutritional content, and, crucially, was far better able to cope with high temperatures. Devaux et al. (2021) suggested that producing true hybrids would have the highest effect over a range of conditions (Table 3).

Controlled-Environment Agriculture

In terms of farming innovations, these would all come under the wider category of "controlled-environment agriculture" which includes everything from a standard unheated greenhouse through to hydroponics (growing without soil, in water-based solutions), aeroponics (growing in air), aquaculture (see Chapter 4), and aquaponics (a hybrid hydroponics system where the nutrition is provided by fish or seafood being produced by aquaculture). Of these categories, hydroponics and aquaponics are likely to provide the most effective future solutions for both tomato and potato productions. For fruit and vegetables, indoor farming (also known as "protected cropping" (Chavan et al. 2022) can be combined with hydroponics and aeroponics, and Despommier (2010, 2014) is an enthusiastic advocate incorporating this into a "vertical farm." O' Sullivan et al. (2020) too make the valid point that if "indoor farming is to realize its potential to increase fruit and vegetable production globally, the diversity of crops that can be grown indoors must increase" (2020, p. 1). Chavan et al. (2022) say that "Indoor vertical farming has a bright future, especially in the wake of the COVID-19 pandemic and is well positioned to increase its share of the global food market, due to its highly efficient production system, reductions in supply chain and logistics costs, potential for automation (minimising handling) and easy access to both labour and consumers" (2022, p. 175). A further advantage of indoor vertical farming is that the process allows for the "stacking" of the plants, whereas outdoor horizontal farming only allows for plants to be planted adjacent to each other. One square kilometre of crop grown outdoors in horizontal farms will occupy slightly more than 1 km^2 of space, but if that same crop area is transposed into an indoor vertical farm, the stacking allows it to use much less floor space.

Both tomatoes and potatoes are suitable for hydroponics and aeroponics, and Huber et al. (2021) demonstrate that "indoor CE systems exhibit higher control of all environmental conditions, including temperature, radiation, spectrum, CO_2 concentration, air velocity, photoperiod, and vapor pressure deficit" all of which are beneficial to improving the performance of tomato plants. Payen et al. (2022) show that "tomato yields in hydroponic systems were significantly greater than tomato yields in soil-based systems" (p. 1), and so this increased yield per plant could lead to a need for fewer plants to produce the same amount of fruit or, more likely, increased production.

Hartinger et al. (2024) note that the cost of seed potatoes "significantly impact potato crop production, especially in tropical and subtropical regions. Aeroponics provides a cost-effective alternative for seed production in these areas" (p. 2). The other advantage of indoor farming/protected cropping is the potential for LED lighting, which can be used in greenhouses to enhance natural light (Ziaei et al. 2024) as well as in environments with no natural light (Dannehl et al. 2021). Dannehl et al. (2021) found that growing tomatoes under LED lights led to higher yield, higher content of the antioxidants lutein and lycopene, and no difference in beta-carotene levels (2021, p. 5). Rahman et al. (2024) tested potatoes under different LED light set-ups and concluded that the use of LEDs "significantly increased the tuber count and size" as well as "higher plant height, branch number and plant biomass" (p). Essentially, growing tomatoes and potatoes under LED lighting seems to produce more, and healthier fruit.

O'Sullivan et al. (2020) neatly sum up the advantages of indoor farming/protected cropping:

> Crops grown indoors are not exposed to extremes of heat or cold, frosts, hail, drought, or flood. Fully enclosed farms physically exclude pests and diseases, resulting in a reduction in the use of fungicides and pesticides. Finally, growing perishable crops closer to consumers can shorten supply chains, reduce transportation and the associated economic and environmental costs of food miles, decrease storage times, and provide consumers with products that have a longer shelf life. (2020, p. 2)

However, the effects of indoor farming are not all positive. At the moment, the majority of vertical indoor farms are powered from the relevant national grid, and most national grids are not 100% fossil fuel free. This means that some of the beneficial impact of this form of farming is tempered by non-renewable energy use. However, as grids go greener, and the feasibility of office block-type buildings embedding renewable energy into their design, the overall sustainability of vertical indoor farming will improve.

A second area of concern is the relationship with pollinators. While it is seen as a positive that moving plants indoors removes the potential for predation by insects and other pests (thus reducing the need for insecticides, pesticides, and fungicides), it can also break the link with pollinating insects. Phillips (2019) reports that commercially produced bees have been used in indoor environments to pollinate plants, but that they need ultraviolet light (UV) in order to help their navigation. UV light does not currently feature in the majority of commercial LED set-ups, meaning the bees can become disorientated. This leads to poor hive health, and "in tomato greenhouses, the populations never do recover when reared under light from electric lamps" (2019).

In conclusion, the future of commercial tomato and potato farming will be profoundly influenced by climate change and advancements in controlled-environment agriculture. As unpredictable weather patterns, intensifying storm systems, and shifting pollinator populations challenge outdoor production, genetic modification and crop breeding will be key tools in creating more resilient varieties. Although public caution may temper the rapid adoption of genetically modified organisms, projects like TomGEM and research on Desiree potatoes show the potential to enhance yields, quality, and stress tolerance.

At the same time, controlled-environment agriculture (CEA) offers a viable strategy for sustaining and expanding production, particularly in an era of resource constraints and growing demand. By enabling year-round cultivation with fewer inputs, indoor farming has the potential to increase yields, reduce pesticide use, and shorten supply chains. However, challenges remain, including the need for cleaner energy sources to power indoor farms and the difficulty of ensuring adequate pollination without natural insect populations.

Overall, while the threat of climate change and evolving consumer expectations will drive significant change, the combined potential of genetic innovations and controlled-environment farming practices provides a promising, if complex, pathway to secure and sustainable tomato and potato production in the coming decades.

Conclusion

The idea of ceasing the commercial production of tomatoes and potatoes is, as are all the chapter in the book, a thought experiment. It does, however, show the connections: agriculture, environment, economy, culture, and justice. Tomatoes and potatoes are more than dietary staples; they are central to livelihoods, ecosystems, and food security globally, and any decision on their future is fraught with complexities and trade-offs.

From an environmental justice perspective, commercial tomato and potato farming is resource intensive, particularly in its use of water and dependence on monoculture practices. Tomatoes rank among the highest water-demanding crops, exacerbating water scarcity in already vulnerable regions. Potatoes, while less water intensive, still require significant irrigation and fertilisers in industrial settings. The transition to more sustainable practices is essential to reduce these environmental burdens. Ceasing production would alleviate some of these pressures, but it would shift the environmental impacts to alternative crops that might not be significantly less resource intensive.

In addition to this is the use of exploited labour forces, whether under the *caporalato* system in Italy or in the "re-education" camps in Xinjiang, which are rife with allegations of forced labour. It would be naïve in the extreme to assume that with the lack of tomato production the Italian *caporale* (and the Organised Criminal networks which feed the system with victims) or the Xinjiang centres would simply cease. They would, of course, simply expand into other markets.

The adoption of sustainable practices such as improved irrigation efficiency, organic farming methods, and crop rotation, and better employment practices could mitigate these concerns without the drastic measure of halting production entirely.

The ecological implications of industrial farming—soil degradation, pesticide use, and biodiversity loss—underscore the urgent need for reform. Monoculture farming of tomatoes and potatoes reduces ecosystem resilience, harms pollinators, and increases pesticide reliance. GM variants offer potential solutions for pest resistance and climate adaptability but raise contested concerns about unintended ecological consequences and the erosion of biodiversity. A cessation of production could potentially allow ecosystems to recover and encourage the revival of indigenous and heritage crop varieties, aligning with species justice principles. However, an alternative approach would be to integrate ecological farming practices, such as polyculture, reduced chemical inputs, and the preservation of genetic diversity, into commercial systems.

The economic ramifications of ceasing commercial production are stark. Tomatoes and potatoes are linchpins of global trade, supporting extensive supply chains and providing income for millions of farmers, particularly smallholders in developing nations. A sudden halt would lead to widespread economic chaos, loss of livelihoods, and increased food insecurity, especially in regions heavily reliant on these crops for subsistence and trade. For a near parallel, consider the serious impact on global wheat markets following the Russian invasion of Ukraine in 2022 (Devadoss and Ridley 2024). Adapting current practices, rather than ceasing production, appears more viable from an economic justice perspective. Investments in sustainable technologies, subsidies for small-scale farmers, and equitable labour protections can mitigate many of the ethical and economic issues associated with industrial farming while maintaining the economic benefits these crops provide.

The future of tomato and potato production lies in innovation. Controlled-environment agriculture (CEA), including hydroponics, vertical farming, and advanced genetic modifications, offers promising pathways to reduce the environmental footprint of these crops. These technologies can enhance water efficiency, reduce pesticide use, and enable year-round production in urban settings, mitigating the pressures of climate change and land degradation. However, CEA is not without challenges. High energy demands, reliance on non-renewable power sources, and potential disruptions to natural pollination systems need to be addressed. Transitioning to cleaner energy and developing technologies that integrate natural pollination processes are critical for the long-term sustainability of these systems.

Overall, we can see that while the cessation of commercial tomato and potato production would have some significant benefits, these would be outweighed by the disruptions. The environmental and ethical issues associated with these crops are not insurmountable and can be addressed through targeted interventions and systemic reforms. Adaptation, rather than cessation, would appear to be the more pragmatic and equitable path forward.

Sources

Abraham, M., and P. Pingali. 2020. Transforming smallholder agriculture to achieve the SDGs. In *The role of smallholder farms in food and nutrition security*, ed. by S. Gomez y Paloma, L. Reisgo, and Louhicji. Springer. https://doi.org/10.1007/978-3-030-42148-9.pdf.

Abushal, L.T., M. Salama, M.M. Essa, and M.W. Qoronfleh. 2021. Agricultural biotechnology: Revealing insights about ethical concerns. *Journal of Biosciences* 46 (3). https://doi.org/10.1007/s12038-021-00203-0.

AGES. 2024. Health for humans, animals & plants: Glycoalkaloids, Österreichische Agentur für Gesundheit und Ernährungssicherheit GmbH. https://www.ages.at/en/human/nutrition-food/residues-contaminants-from-a-to-z/glycoalkaloids.

Aguiar, J.M.R.B.V., R.C.F. Nocelli, M. Giurfa, and F.S. Nascimento. 2023. Neonicotinoid effects on tropical bees: Imidacloprid impairs innate appetitive responsiveness, learning and memory in the stingless bee Melipona quadrifasciata. *The Science of the Total Environment*, 2023-06 877: 162859–162859, Article 162859. https://doi.org/10.1016/j.scitotenv.2023.162859.

Aktar, W.M., D. Sengupta, and A. Chowdhury. 2009. Impact of pesticides use in agriculture: Their benefits and hazards. *Interdisciplinary Toxicology* 2 (1): 1–12. https://doi.org/10.2478/v10102-009-0001-7.

Allan, J.A. 1993. Fortunately there are substitutes for water otherwise our hydropolitical futures would be impossible. In *ODA, priorities for water resources allocation and management*, 13–26. London: ODA.

Amnesty International. 2012. *Exploited labour: Migrant workers in Italy's agricultural sector*. London: Amnesty International Publications.

Aznar-Sánchez, J.A., L.J. Belmonte-Ureña, and J.J. Tapia-León. 2014. The industrial agriculture A 'model for modernization' from Almería?. In *Seasonal workers in Mediterranean agriculture: The social costs of eating fresh*, ed. by G. Gertel, and S.R. Sippell. London: Routledge.

Bade, A.S., and V.K. Bhamare. 2024. Insecticidal resistance in Helicoverpa armigera (Hubner) infesting chickpea. *Indian Journal of Entomology* 86 (1): 204–207. https://doi.org/10.55446/IJE.2023.1348.

Bagnardi, F., D'Onofrio, G., and Greco, L., 2022. The state in Chains: Public Policies against adverse incorporation in Southern Italian production networks. *Globalizations*, 19 (1): 34–58

BBC Eye. 2024. Blood on the shelves: The secrets of Xinjiang's tomato industry. BBC Eye. https://www.youtube.com/watch?v=dtv_1eafJn4.

Bearth, A., G. Kaptan, and S.H. Kessler. 2022. Genome-edited *versus* genetically-modified tomatoes: An experiment on people's perceptions and acceptance of food biotechnology in the UK and Switzerland. *Agriculture and Human Values* 39: 1117–1131. https://doi.org/10.1007/s10460-022-10311-8.

Bita, C.E., and T. Gerats. 2013. Plant tolerance to high temperature in a changing environment: Scientific fundamentals and production of heat stress-tolerant crops. *Frontiers in Plant Science* 4: 273. https://doi.org/10.3389/fpls.2013.00273.

Bogužas, V., L. Skinulienė, L.M. Butkevičienė, V. Steponavičienė, E. Petrauskas, and N. Maršalkienė. 2022. The effect of monoculture, crop rotation combinations, and continuous bare fallow on soil CO_2 emissions, earthworms, and productivity of winter rye after a 50-year period. *Plants (Basel)* 11 (3): 431. https://doi.org/10.3390/plants11030431.

Brooker, R.W., R. Hewison, C. Mitchell, A.C. Newton, R.J. Pakeman, C. Schöb, C., and A.J. Karley. 2021. Does crop genetic diversity support positive biodiversity effects under experimental drought? *Basic and Applied Ecology* 56: 431–445. https://doi.org/10.1016/j.baae.2021.05.001.

Burauel, P., and F. Baßmann. 2005. Soils as filter and buffer for pesticides—Experimental concepts to understand soil functions. *Environmental Pollution* 133 (1): 11–16. https://doi.org/10.1016/j.envpol.2004.04.011.

Castro, A.J., M.D. López-Rodríguez, C. Cynthia Giagnocavo, M. Gimenez, L. Céspedes, A. La Calle, M. Gallardo, P. Pumares, J. Cabello, E. Rodríguez, D. Uclés, S. Parra, J. Casas, F. Rodríguez, J.S. Fernandez-Prados, M. Expósito-Granados, B.E. Murillo-López, L.M. Vasquez, and D.L. Valera. 2019. Six collective challenges for sustainability of Almería greenhouse horticulture.

International Journal of Environmental Research and Public Health 16: 4097. https://doi.org/10.3390/ijerph16214097.

Chavan, S.G., Z.-H. Chen, O. Ghannoum, C.I. Cazzonelli, and D.T. Tissue. 2022. Current technologies and target crops: A review on Australian protected cropping. *Crops* 2 (2): 172–185. https://doi.org/10.3390/crops2020013.

Clapp, J. 2015. Mega-mergers on the menu: Corporate concentration and the politics of sustainability in the global food system. *Global Environmental Politics* 18: 2. https://doi.org/10.1162/glep_a_00454.

Dannehl, D., T. Schwend, D. Veit, and U. Schmidt. 2021. Increase of yield, lycopene, and lutein content in tomatoes grown under continuous PAR spectrum LED lighting. *Frontiers in Plant Science* 12. https://doi.org/10.3389/fpls.2021.611236.

Despommier, D. 2010. *The vertical farm: Feeding the world in the 21st century*. London: Picador.

Despommier, D. 2014. Vertical farms in horticulture. In: *Encyclopedia of food and agricultural ethics*, ed. by P.B. Thompson and D.M. Kaplan. Dordrecht: Springer. https://doi.org/10.1007/978-94-007-0929-4_88.

Devadoss, S., and W. Ridley. 2024. Impacts of the Russian invasion of Ukraine on the global wheat market. *World Development* 173: 106396. https://doi.org/10.1016/j.worlddev.2023.106396.

Devaux, A., J-P. Goffart, P. Kromann, J. Andrade-Piedra, V. Polar, and G. Hareau. 2021. The potato of the future: Opportunities and challenges in sustainable agri-food systems. *Potato Research* 64: 681–720. https://doi.org/10.1007/s11540-021-09501-4.

Dongyu, Q. 2022. Role and potential of potato in global food security. Food and Agricultural Organization of the United Nations. https://openknowledge.fao.org/server/api/core/bitstreams/92ca0047-7f4c-4ac3-bd16-0153662275b0/.

EFSA. 2024. Genetically modified organisms. European Food Safety Authority. https://www.efsa.europa.eu/en/topics/genetically-modified-organisms.

Evans, D.L., N. Falagán, C.A. Hardman, S. Kourmpetli, L. Liu, B.R. Mead, and J.A.C. Davies. 2022. Ecosystem service delivery by urban agriculture and green infrastructure—A systematic review. *Ecosystem Services* 54: 101405, Article 101405. https://doi.org/10.1016/j.ecoser.2022.101405.

FAO. 2024. Biotechnology: FAO's role in biotechnology. UN Food and Agriculture Organization. https://www.fao.org/biotechnology/en.

FAOSTAT. 2022. Tomato, food and agricultural organization of the United Nations. https://www.fao.org/faostat/en/#home.

FAOSTAT. 2023. FAOSTAT analytical brief 79: Agricultural production statistics 2000–2022. Food and Agricultural Organization of the United Nations. https://openknowledge.fao.org/server/api/core/bitstreams/fba4ef43-422c-4d73-886e-3016ff47df52/content.

Foley, J.A., N. Ramankutty, K.A. Brauman, E.S. Cassidy, J.S. Gerber, and M. Johnston. 2011. Nature 478: 337–342. https://doi.org/10.1038/nature10452.

Friedman, M., and G.M. McDonald. 1997. Potato Glycoalkaloids: Chemistry, analysis, safety, and plant physiology. *Critical Reviews in Plant Sciences* 16 (1): SS-132. https://doi.org/10.1080/07352689709701946.

Gold, A., W. Fung, S. Gabbard, and D. Carroll. 2022. Research report number 16: Findings from the national agricultural workers survey (NAWS) 2019–2020: A demographic and employment profile of United States Farmworkers. JBS International/US Department of Labor. https://www.dol.gov/sites/dolgov/files/ETA/publications/ETAOP2022-16_NAWS_Research_Report_16_508c.pdf.

Goulson, D. 2013. An overview of the environmental risks posed by neonicotinoid insecticides. *Journal of Applied Ecology* 50 (4): 977–987. https://doi.org/10.1111/1365-2664.12111.

Guinet, M., G. Adeux, S. Cordeau, E. Courson, R. Nandillon, Y. Zhang, and N. Munier-Jolain. 2023. Fostering temporal crop diversification to reduce pesticide use. *Nature Communications* 14: Article number 7416. https://doi.org/10.1038/s41467-023-43234-x.

Gupta, H., U. Deeksha, and S.G. Reddy. 2023. Insecticidal and detoxification enzyme inhibition activities of essential oils for the control of pulse beetle, Callosobruchus maculatus (F.) and Callosobruchus chinensis (L.) (Coleoptera: Bruchidae). *Molecules (Basel, Switzerland)* 28 (2): 492. https://doi.org/10.3390/molecules28020492.

GWR. 2024. Heaviest tomato. Guinness World Records. https://www.guinnessworldrecords.com/world-records/heaviest-tomato.

Hajjar, R., D.I. Jarvis, and B. Gemmill-Herren. 2008. The utility of crop genetic diversity in maintaining ecosystem services. *Agriculture, Ecosystems & Environment* 123 (4): 261–270. https://doi.org/10.1016/j.agee.2007.08.003.

Hartinger, S.C., C.K. de Matos, R. Moccellin, C.M.D.R.F. Faria, and J. Kawakami. 2024. Plant growth–promoting microorganisms: Performance of potato cultivars in aeroponics. *Potato Research*. https://doi.org/10.1007/s11540-024-09813-1.

Haverkort, A.J., and P.C. Struik. 2015. Yield levels of potato crops: Recent achievements and future prospects. *Field Crops Research* 182: 76–85. https://doi.org/10.1016/j.fcr.2015.06.002.

Hill, W.G. 2013. Selective breeding. In *Brenner's Encyclopedia of genetics*, ed. by S. Maloy and K. Hughes, 371–373. https://doi.org/10.1016/B978-0-12-374984-0.01390-5.

Hoekstra, A.Y., and Champagain, A.K., Water footprints of nations: Water use by people as a function of their consumption pattern, Water Resource Management (2007) 21: 35–48. https://doi.org/10.1007/s11269-006-9039-x.

Howard, P.H. 2009. Visualizing consolidation in the global seed industry: 1996–2008. *Sustainability* 1 (4): 1266–1287. https://doi.org/10.3390/su1041266.

HRW. 2012. Cultivating fear: The vulnerability of immigrant farmworkers in the US to sexual violence and sexual harassment. Human Rights Watch. https://www.hrw.org/report/2012/05/15/cultivating-fear/vulnerability-immigrant-farmworkers-us-sexual-violence-and.

Huber, B.M., F.J. Louws, and R. Hernández. 2021. Impact of different daily light integrals and carbon dioxide concentrations on the growth, morphology, and production efficiency of tomato seedlings. *Frontiers in Plant Science* 3 (12): 615853. https://doi.org/10.3389/fpls.2021.

Indexbox. 2024. World—Potato—Market analysis, forecast, size, trends and insights. https://www.indexbox.io/store/world-potato-market-report-analysis-and-forecast-to-2020/.

IPCC. 2019. *Climate change and land: An IPCC special report on climate change, desertification, land degradation, sustainable land management, food security, and greenhouse gas fluxes in terrestrial ecosystems* [P.R. Shukla, J. Skea, E. Calvo Buendia, V. Masson-Delmotte, H.-O. Pörtner, D. C. Roberts, P. Zhai, R. Slade, S. Connors, R. van Diemen, M. Ferrat, E. Haughey, S. Luz, S. Neogi, M. Pathak, J. Petzold, J. Portugal Pereira, P. Vyas, E. Huntley, K. Kissick, M. Belkacemi, J. Malley, (eds.)]. https://www.ipcc.ch/srccl/.

Jørs, E., R.C. Morant, G.C. Aguilar, O. Huici, F. Lander, J. Baelum, and F. Konradsen. 2006. Occupational pesticide intoxications among farmers in Bolivia: A cross-sectional study. *Environmental Health* 5: 10. https://doi.org/10.1186/1476-069X-5-10.

Keese, P. 2008. Risks from GMOs due to horizontal gene transfer. *Environmental Biosafety Research* 7: 123–149. https://doi.org/10.1051/ebr:2008014.

Kennedy, I. 1999. *Genetically modified crops: The ethical and social issues.* London: Nuffield Council on Bioethics.

Kesavachandran, C.N., M. Fareed, M.J. Pathak, V. Bihari, N.. Mathur, and A.K. Srivastava. 2009. Adverse health effects of pesticides in agrarian populations of developing countries. In *Reviews of environmental contamination and toxicology*, ed. by D.M. Whitacre, 200, 33–52. https://doi.org/10.1007/978-1-4419-0028-9_2.

Kim, K.-H., E. Kabir, and S.A. Jahan. 2017. Exposure to pesticides and the associated human health effects. *Science of the Total Environment* 575: 525–535. https://doi.org/10.1016/j.scitotenv.2016.09.009.

Kramer, M.G., and K. Redenbaugh. 1994. Commercialization of a tomato with an antisense polygalacturonase gene: The FLAVR SAVR™ tomato story. *Euphytica* 79: 293–297. https://doi.org/10.1007/BF00022530.

Lal, R. 2001. Soil degradation by erosion. *Land Degradation & Development* 12: 519–539. https://doi.org/10.1002/ldr.472.

Lal, R. 2009. Soil degradation as a reason for inadequate human nutrition. *Food Security.* 1: 45–57. https://doi.org/10.1007/s12571-009-0009-z.

Littletom. 2016. Tomberry Pareltomaten: Informatie, Littletom BV. https://www.littletom.nl/tomberry

Martineau, B. 2001. *First fruit: The creation of the Flavr Savr™ tomato and the birth of biotech foods*. New York: McGraw-Hill.

Meacham-Hensold, K., A.P. Cavanagh, P. Sorensen, P.F. South, J. Fowler, R. Boyd, J. Jeong, S. Burgess, S. Stutz, Ryan N. Dilger, M. Lee, N. Ferrari, J. Larkin, and D.R. Ort. 2024. Shortcutting photorespiration protects potato photosynthesis and tuber yield against heatwave stress. *Global Change Biology* 30 (12): e17595. https://doi.org/10.1111/gcb.17595.

Mostafalou, S., and M. Abdollahi. 2017. Pesticides: An update of human exposure and toxicity. *Archives of Toxicology* 91 (2): 549–599. https://doi.org/10.1007/s00204-016-1849-x.

Naresh Kumar, S., P.M. Govindakrishnan, D.N. Swarooparani, C. Nitin, J. Surabhi, P.K. Aggarwal. 2015. Assessment of impact of climate change on potato and potential adaptation gains in the Indo-Gangetic Plains of India. *International Journal of Plant Production* 9: 151–169.

NASA. 2022. Earth observatory: Almería's sea of greenhouses. https://earthobservatory.nasa.gov/images/150070/almerias-sea-of-greenhouses.

NBS. 2024. National data. National Bureau of Statistics China. https://data.stats.gov.cn/english/easyquery.htm.

NHS. 2024. Overview: Vitamins and minerals. National Health Service. https://www.nhs.uk/conditions/vitamins-and-minerals/.

OCHCR. 2022. OHCHR assessment of human rights concerns in the Xinjiang Uyghur Autonomous Region, People's Republic of China, Office of the High Commissioner for Human Rights. https://www.ohchr.org/en/documents/country-reports/ohchr-assessment-human-rights-concerns-xinjiang-uyghur-autonomous-region.

ODNI. 2021. The future of water: Water insecurity threatening global economic growth. Political Stability Office of the Director of National Security. https://www.dni.gov/index.php/gt2040-home/gt2040-deeper-looks/future-of-wate.

Odong, L.-M. 2024. Reconciling the incongruence between the Cartagena protocol on biosafety to the convention on biological diversity and the GATT/WTO rules. *International Environmental Agreements: Politics, Law and Economics*. https://doi.org/10.1007/s10784-024-09649-7.

Ollerton, J., H. Erenler, M. Edwards, and R. Crockett. 2014. Extinctions of aculeate pollinators in Britain and the role of large-scale agricultural changes. *Science* 346: 1360–1362. https://doi.org/10.1126/science.1257259.

O'Sullivan, C.A., C.L. McIntyre, I.B. Dry, S.M. Hani, Z. Hochman, and G.D. Bonnett. 2020. Vertical farms bear fruit. *Nature Biotechnology* 38 (2). https://doi.org/10.1038/s41587-019-0400-z.

Patel, R. 2009. 2009, What does food sovereignty look like? *The Journal of Peasant Studies* 36 (3): 663–706. https://doi.org/10.1080/03066150903143079.

Payen, F. T., Evans, D. L., Falagán, N., Hardman, C. A., Kourmpetli, S., Liu, L., et al. (2022). How much food can we grow in urban areas? Food production and crop yields of urban agriculture: A meta-analysis. *Earth's Future*, 10, e2022EF002748. https://doi.org/10.1029/2022EF002748.

Perrotta, D. 2019. Processing tomatoes in the era of the retailing revolution. Mechanization and migrant labour in northern and southern Italy. In *Migration and agriculture: Mobility and change in the Mediterranean area*, ed. by A. Corrado, C. de Castro, and D. Perrotta. London: Routledge.

Phillips, B. 2019. Bumble bees in greenhouse vegetable production. Michigan State University Extension, 12 April. https://www.canr.msu.edu/news/bumble-bees-in-greenhouse-vegetable-production.

Pizetta C.S.R., W.R. Ribeiro, A.L. Ferreira, M.M. da Costa, K. Bonfim, P.V. Pinheiro, G.B. Cabral, and F.J.L. Aragão. 2022. RNA interference-mediated tolerance to whitefly (Bemisia tabaci) in genetically engineered tomato. *Plant Cell, Tissue, and Organ Culture; Dordrecht* 148 (2): 281–291. https://doi.org/10.1007/s11240-021-02185-1.

Postel, S.L., G.C. Daily, and P.R. Ehrlich. 1996. Human appropriation of renewable fresh water. *Science* 271 (5250): 785–788. https://doi.org/10.1126/science.271.5250.785.

Prasad, A., N. Sharma, O. Chirom, and M. Prasad. 2022. The sly-miR166-SlyHB module acts as a susceptibility factor during ToLCNDV infection. *Theoretical and Applied Genetics* 135 (1): 233–242. https://doi.org/10.1007/s00122-021-03962-4.

Rahman, M.H., M.J. Islam, H.M. Umma, B-R. Ryu, J-D. Lim, M.O.K. Azad, E.J. Cheong, and Y-S. Lim. 2024. The growth and tuber yield of potatoes (*Solanum tuberosum* L.) under varying LED light spectrums in controlled greenhouse conditions. *Horticulturae* 10 (3): 254. https://doi.org/10.3390/horticulturae10030254.

Reuters. 2024. Spain to legalise about 300,000 undocumented immigrants per year. Reuters, 19 November. https://www.reuters.com/world/europe/spain-regularise-about-300000-undocumented-migrants-per-year-2024-11-19/.

RHS. 2024. Glasshouse Whitefly. Royal Horticultural Society. https://www.rhs.org.uk/biodiversity/glasshouse-whitefly.

Ritchie, H., Rosado, P. and Roser, M. 2024. Environmental Impacts of Food Production, *Our World in Data*, https://ourworldindata.org/environmental-impacts-of-food.

Royal Society. 2024. Genetically modified (GM) plants: questions and answers. Royal Society. https://royalsociety.org/news-resources/projects/gm-plants/.

Sánchez-Bayo, F., and K.A.G. Wyckhuys. 2019. Worldwide decline of the entomofauna: A review of its drivers. *Biological Conservation* 232: 8–27. https://doi.org/10.1016/j.biocon.2019.01.020.

Song, Y., X. Liu, K. Zhao, R. Ma, W. Wu, Y. Zhang, L. Duan, X. Li, H. Xu, M. Cheng, B. Qin, and Z. Qi. 2024. A new endophytic Penicillium oxalicum with aphicidal activity and its infection mechanism. *Pest Management Science* 80 (11): 5706–5717. https://doi.org/10.1002/ps.8288.

Statistics Netherlands. 2024. Netherlands is second-largest exporter and importer of goods in EU. Statistics Netherlands/Centraal Bureau voor de Statistiek. https://www.cbs.nl/en-gb/news/2024/17/netherlands-is-second-largest-exporter-and-importer-of-goods-in-eu.

Stice, C.P., H. Xia, and X.D. Wang. 2018. Tomato lycopene prevention of alcoholic fatty liver disease and hepatocellular carcinoma development. *Chronic Diseases and Translational Medicine* 4 (4): 211–224. https://doi.org/10.1016/j.cdtm.2018.11.001.

Straits Research. 2023. Tomato market size, share & trends analysis report by product (sauces, paste, canned tomatoes, ketchup, juice, others) and by region (North America, Europe, APAC, Middle East and Africa, LATAM) forecasts, 2023–2031. https://straitsresearch.com/report/tomato-market.

TomGEM. 2021. Saving fruit quality for tomorrow. TomGEM Newsletter. https://tomgem.eu/fileadmin/tomgem/downloads/TomGEM_newsletter.pdf.

TrendEconomy. 2024. Potatoes: Imports and exports: 2023. TrendEconomy. https://trendeconomy.com/data/commodity_h2/0701.

Truelove, R.N., S.C. Lellyett, A.I. Issaka, and S. Huda. 2023. Agricultural value chains in developing economies: A theoretical framework. In *Sustainable food value chain development: Perspectives from developing and emerging economies*, ed. by S.A. Narule, and S.P. Raj. Singapore: Springer.

Tscharntke, T., A.M. Klein, A. Kruess, I. Steffan-Dewenter, and C. Thies. 2005. Landscape perspectives on agricultural intensification and biodiversity—Ecosystem service management. *Ecology Letters* 8: 857–874. https://doi.org/10.1111/j.1461-0248.2005.00782.x.

Tunio, M.H., Gao, J., Shaikh, S.A., Lakhiar, I.A., Qureshi, W.A., 2020. Potato production in aeroponics: An emerging food growing system in sustainable agriculture for food security, *Chilean Journal of Agricultural Research*; Chillán Vol. 80, Iss. 1, (Jan-Mar 2020): 118–132. https://doi.org/10.4067/S0718-58392020000100118.

Waltz, E. 2015a. USDA approves next-generation GM potato. *Nature Biotechnology* 33 (1): 12–13. https://doi.org/10.1038/nbt0115-12.

Waltz, E. 2015b. Nonbrowning GM apple cleared for market. *Nature Biotechnology* 33 (4): 326–327. https://doi.org/10.1038/nbt0415-326c.
Wang, N., P. Reidsma, A.A. Pronk, A.J.W. de Wit, and M.K. van Ittersum. 2018. Can potato add to China's food self-sufficiency? *The Scope for Increasing Potato Production in China, European Journal of Agronomy* 101: 20–29. https://doi.org/10.1016/j.eja.2018.07.002.
Wang, Z., H. Liu, F. Zeng, Y. Yang, D. Xu, Y. Zhao, X. Liu, L. Kaur, G. Liu, and J. Singh. 2023. Potato processing industry in China: Current scenario. *Future Trends and Global Impact, Potato Research* 66: 543–562. https://doi.org/10.1007/s11540-022-09588-3.
White, R. 2014. *Environmental harm: An eco-justice perspective*. Bristol: Bristol University Press.
World Bank. 2024. Agriculture and food: Overview. World Bank. https://www.worldbank.org/en/topic/agriculture/overview.
Zheng, S., L. Yu, and H. Fu. 2024. Has rural e-commerce increased potato farmers' income? Evidence from the potato home of China. *Potato Research* 67: 15–35. https://doi.org/10.1007/s11540-023-09614-y.
Ziaei, N., M. Talebi, B.E.S. Tabatabaei, M.R. Sabzalian, and M. Soleimani. 2024. Intra-canopy LED lighting outperformed top LED lighting in improving tomato yield and expression of the genes responsible for lycopene, phytoene and vitamin C synthesis. *Scientific Reports* 14:19043. https://doi.org/10.1038/s41598-024-69210-z.

CHAPTER 4

Fish and Seafood

Abstract This chapter analyses the environmental, social, and ethical challenges associated with the farming of key aquaculture species: tiger prawns, king prawns, Atlantic salmon, and tilapia. These species dominate global aquaculture, providing a crucial source of protein and economic livelihoods but also contributing to significant environmental degradation and social injustices. The chapter outlines the environmental impacts of aquaculture, including water pollution, habitat destruction, and the overuse of antibiotics, which exacerbate biodiversity loss and contribute to the emergence of antibiotic-resistant pathogens. Social issues such as labour exploitation, including instances of human trafficking and unsafe working conditions, are examined in the context of prawn and fish farming industries in regions including Southeast Asia, South America, and Africa. The ethical implications of farming sentient creatures in intensive, high-density systems are also explored, with attention to welfare concerns and inadequate regulatory frameworks. The chapter concludes by highlighting the need for sustainable aquaculture practices and innovative alternatives to balance global food security, economic benefits, and ecological resilience.

Keywords Aquaculture · Antibiotic resistance · Biodiversity loss · Labour exploitation · Sustainable food systems

© The Author(s), under exclusive license to Springer Nature Switzerland AG 2025

S. Sneddon, *Food, Environmental Degradation and Injustices*, Palgrave Studies in Green Criminology,
https://doi.org/10.1007/978-3-031-87900-5_4

Introduction

Farmed prawns and farmed fish will be the focus for this chapter. It is of course more difficult to discuss aggregated species such as "prawn" and "fish" when contrasted with the specific discussions in previous chapters. For that reason, although the environmental impacts (Ritchie et al. 2024) are calculated on generic species, this chapter will focus on tiger prawns (*Penaeus monodon*), king prawns (Lito*penaeus vannamei*, also called whiteleg shrimp), Atlantic salmon (*Salmo* salar) and tilapia,[1] which are the most commonly farmed species across the world, and which dominate aquaculture production globally.

Farmed prawns and fish are vital to the global food industry, representing a rapidly growing sector within aquaculture. The chapter considers the key issues surrounding the farming of the four names species, although on occasion it has bene necessary to aggregate the discussion. These species are cultivated to meet the escalating demand for seafood, driven by factors including population growth, urbanisation, and the increasing perception of seafood as a healthier protein alternative. While aquaculture presents an opportunity to alleviate overfishing and support food security, it also raises complex environmental, social, and ethical challenges.

The global expansion of aquaculture is remarkable. Between 1990 and 2020, aquaculture grew at an annual rate of 7.5%, far outpacing the growth of wild-capture fisheries. By 2022, aquaculture surpassed wild fisheries in global seafood production (FAO 2024a), illustrating its centrality to modern food systems. This growth, however, comes with significant trade-offs. Intensive farming practices, high stocking densities, and reliance on chemicals and antibiotics have made aquaculture a contentious industry. The sections below will address some of these complexities, focusing on the intertwined aspects of environmental, ecological, and species justice, as well as considering the economic angle.

As with all aquacultures, the farming of tiger prawns and king prawns has experienced exponential growth over recent decades (Sherratt 2020) due to increasing global demand for seafood. Prawn aquaculture, particularly in regions such as Southeast Asia and Latin America, contributes significantly to local economies through employment and exports and is

[1] Tilapia is a generic name for many distinct species, so has not had a single Latin name attributed to it here.

a multibillion-dollar industry (WWF 2024). However, this industry has been criticised for severe environmental degradation and social inequalities, which disproportionately affect poor and marginalised communities.

Although the impacts of salmon and tilapia farming are very similar in nature, the regions impacts are quite different. Norway and Chile are by far the two largest producers of farmed salmon (Carrasco-Bahamonde and Casellas 2024), but the practice is also significant in other countries (for example Canada and Scotland). Tilapia, on the other hand tilapia is cultivated extensively in tropical and subtropical countries, especially China (the world's largest producer), Egypt, and Indonesia (Yuan et al. 2020). Though aquaculture has brought economic benefits and provided a source of affordable protein, the industry is also associated with severe environmental degradation and social inequalities. Water pollution, the overuse of chemicals, and labour exploitation in fish farms have significant human health and social justice implications, especially for workers, local communities, and consumers.

Tiger prawns and king prawns are pivotal to aquaculture, particularly in Southeast Asia and Latin America, where their farming supports local economies. However, this prosperity often comes at an excessive cost. Mangrove deforestation, coastal habitat destruction, and pollution caused by shrimp farming have far-reaching ecological consequences. Moreover, social inequalities, including labour exploitation and inadequate wages, disproportionately affect marginalised communities reliant on aquaculture for their livelihoods. These environmental and social injustices are mirrored in the farming of salmon and tilapia, which are cultivated primarily in colder climates like Norway and Chile for salmon and tropical regions like Egypt and China for tilapia.

The environmental justice implications of aquaculture extend beyond localised pollution and habitat destruction. The overuse of antibiotics in fish and prawn farming contributes to the global crisis of antibiotic resistance, posing a significant public health threat. Simultaneously, water degradation caused by nutrient run-off and chemical pollutants exacerbates ecological harm, reducing biodiversity and creating hypoxic zones in aquatic ecosystems. The chapter also explores how these practices intersect with global sustainability goals, such as the United Nations' Sustainable Development Goal 14, which calls for the conservation and sustainable use of oceans, seas, and marine resources. Ceasing the global market for farmed prawns and farmed fish would therefore have both beneficial and detrimental impacts.

The first section of the chapter will consider this from an environmental justice perspective, highlighting the key human health concerns.

Environmental Justice

There are several environmental justice issues associated with the farming of prawns and fish. Broadly speaking, these can be separated into human health risks—those associated with consumption, those associated with the environmental pollution caused by these types of aquaculture, and those associated with the farming processes themselves—and the social impacts which emerge as a result of this industry.

Water Degradation, Consumption, and Human Health Risks

The farming of tiger and king prawns causes significant pollution of water, as the "wastewater collected from shrimp aquaculture farms are rich in nitrogen, phosphorous, salts, heavy metals and many other toxic substances" (Iber and Kasan 2021). Partly these polluting substances come from the faeces of the prawns, but also from the pesticides, fertilisers, antibiotics, and antimicrobial chemicals which are added to the water to ensure maximum yield. When the water is periodically discharged into nearby rivers, estuaries, or coastal waters, it can lead to water pollution, including eutrophication, harmful algal blooms, and oxygen depletion (Naylor et al. 2000). The same issues occur with salmon and tilapia farming as these also take place predominantly in pens located in coastal waters (or fjords in Norway). Again, fish faeces, uneaten food, and excess chemicals accumulate in the water around the fish farms. This waste increases nutrient loads in local water systems, leading to eutrophication, harmful algal blooms, and oxygen depletion, all of which can have significant environmental and public health consequences (Burridge et al. 2010).

As was demonstrated in Chap. 2 (Meat), eutrophication and oxygen depletion lead to hypoxia and anoxia in rivers and waterways, and this means both that the water will not support any non-human life, but also that it is not fit for human consumption. In coastal communities, the degradation of water and contamination of local water sources has significant and long-term consequences. The United Nations' Sustainable Development Goal 6 is "Ensure access to water and sanitation for all", and further make the point that "access to safe water, sanitation and hygiene is the most basic human need for health and well-being" (UN 2024).

The groups of people who are most immediately impacted by this polluted water are those who are employed to work in the farms. This is not a new phenomenon, and despite the shrimp and prawn industry making claims about improved use of chemicals and antibiotics in the farming process (see, for example Towers 2012), Holmström et al. (2003) were raising concerns more than two decades ago. As Heal et al. (2021) note, the threat to human health has not disappeared. Part of the reason for the human health risks is the lack of effective protective equipment. As we will see shortly, prawn, shrimp, and fish farmers are often poorly paid and exploited.

These farms pose a potential risk to the wider population. Antibiotics are widely used in aquaculture to keep the fish and prawn stocks healthy and free from diseases. It is impossible, however, to accurately administer antibiotics to a pen filled with up to 200,000 salmon (MOWI 2023) and so it is commonplace to over-use, rather than under-use antibiotics—although "precise data is often difficult to obtain since usage is widely unregulated and indiscriminate, with antibiotics freely available 'over the counter' and little capacity for enforcing regulations" (Thornber et al. 2019). This results in unused antibiotics entering the water and the wider food web, leading to a distinct possibility of antibiotic-resistant diseases emerging (Cabello 2006). The further risk is that these pathogens become zoonotic, that is they cross the species boundary into humans, in the same way as coronavirus SARS-CoV-2 (COVID-19).

In addition to the antibiotic risk, Ortiz-Cañavate et al. (2019) have found that run-off from fungicides, even if applied at correct levels, can lead to the creation of cyanobacteria, which is a major cause of algal blooms. Uddin et al. (2016), Chang et al. (2019), Wang et al. (2023), and others have all expressed concern at the health implications of the use of organophosphate insecticides in aquaculture. Uddin et al. (2016) were primarily concerned with the health of the aquatic life more widely, arguing that the drift of these pesticides meant that their impact was felt beyond the confines of the farm itself. Chang et al. (2019) demonstrated that residues of phthalates (a chemical primarily used to soften plastics) and insecticides were detectable in farmed prawns. The levels were relatively low, and although the immediate risk to human health was "negligible", they called for "continual monitoring of the residues in shrimps [...] for further assessment of possible effects on human health" (p. 140).

Wang et al. (2023) focused on organophosphate ethers (OPE, a type of chemical used as a flame retardant) and concluded that while "there is a low, but not negligible, risk to human health from OPEs", "the long-term and sustained consumption of OPE-contaminated diets can increase the risk of exposure" (p. 16). The UN Food and Agriculture Organization estimated that on average humans consume just over 20 kg of fish and seafood per year, and this ranges from an average of 5.3 kg in low-income countries, through to "12.5 kg in lower-middle-income countries, 30.6 kg in upper-middle-income countries and 26.7 kg in high-income countries" (FAO 2024a, p. 75). Even though OPEs are bioaccumulative in humans (our bodies store them, rather than processing and discarding them), and the overall current risk is low, and the FAO figures suggest the risk might be higher in upper-middle-income and high-income countries where the most fish and seafood is consumed.

Potential harm to humans is not limited to concerns over consumption of OPEs, however. Williams (2023) reports that as a result of the "chemicals, including pesticides (usually organo-phosphorous pesticides), disinfectants" (2023, p. 293), which are harmful if inhaled or absorbed through the skin, workers on fish farms and shellfish farms are exposed to risks of respirational illnesses such as asthma and COPD (chronic obstructive pulmonary disease), as well as contact dermatitis and other skin irritations. Fish and Prawns may also contain several species of *mycobacteria*, some of which are zoonotic, and can transfer to humans. Infection of fish farm workers with these strains of mycobacteria can cause a range of health issues, from the minor and temporary (for example skin lesions) to more serious problems such as necrosis (the death of tissue, and a precursor to gangrene), arthritis, osteomyelitis, and bronchitis (Ziarati et al. 2022).

Having considered the potential health implications from consuming and working with farmed fish and prawns, we must now turn our attention to the second major injustice associated with the trade—namely the exploitation and poor treatment of workers in the industry.

Labour Exploitation and Health Impacts

The human health impacts of prawn farming are not limited to environmental pollution. Stringer et al. (2022) concluded that there was significant interaction in the Thai fishing industry between unscrupulous companies and existing human trafficking or modern slavery networks, resulting in "physical and sexual abuse [...] coercion, debt bondage, [...] slave-like conditions onboard vessels, [and] torture and execution..." (2022, p. 6). Concerns about migrant workers, whatever their status (legal migrants, illegal migrants, victims of trafficking) are always greater, because if a person lacks legal status there is an increased likelihood that they will lack access to healthcare provisions, employment protection, and recourse to legal services. In 2019, the UN Special Rapporteur on the Right to Food said that:

> Most of the 120 million people working in the fisheries sector - often exposed to dangerous conditions and working up to 20 hours a day - don't earn a living wage. They therefore fail to fulfil their families' basic needs including food, clothing, housing, education and healthcare (OHCHR 2019)

In 2023, the UN Special Rapporteur on the issue of human rights obligations relating to the enjoyment of a safe, clean, healthy, and sustainable environment visited Chile and, after assessing the country's billion-dollar farmed salmon industry, recommended a "moratorium on further expansion of salmon aquaculture pending an independent scientific review of the adverse environmental impacts" (2024, Recommendation (h)(iv), p. 19). Greenpeace (2021) calculate that the Chilean salmon industry had grown at a rate of 117% a year since the 1990s, and that this is causing damage to ecosystems which have been used by Indigenous communities for millennia. Elsewhere, the pressure group Ecoceanos uses official Chilean data to report that nearly 80 divers died in the country's salmon farms in the decade to 2023 (Ecoceanos 2023).

It is not just Southeast Asia and South America which have faced problems for their treatment of workers. In the United States, farmed tilapia was for a while farmed by prisoners in Colorado. The fish farms were operated by Colorado Correctional Industries, a division of the Colorado Department of Corrections, which was "created by the Colorado Legislature in 1976, with the purpose of creating profitable industries inside prison facilities to provide work opportunities for inmates while they are incarcerated" (CCI 2024). The revelation in 2015 that the workers were paid far less than minimum wage (Wood 2015) caused the main client (Whole Foods) to terminate their tilapia relationship with CCI in 2016 (Romero 2022). The report was confirmed by the Colorado State Legislature in 2017, which reported that "Inmates work an average of 34 hours a week earning $0.84/day" (LCS 2017). The Colorado State Law was changed in 2022, and state prisoners are now entitled to minimum wage. The end result is that tilapia production became more expensive in the United States, and sales have fallen as a result.

The section has highlighted the environmental injustices associated with the farming of prawns, salmon, and tilapia, focusing on the health and other impacts on humans of water pollution, consumption, and labour exploitation. The environmental impact includes severe water contamination from waste and chemicals used in aquaculture. Farms for tiger prawns, king prawns, salmon, and tilapia release nitrogen, phosphorus, heavy metals, and antibiotics into surrounding waters, leading to eutrophication, harmful algal blooms, and oxygen depletion, which damage ecosystems and render water unsafe for consumption. This degradation poses health risks to nearby communities and to farm workers who lack proper protective equipment, exposing them to respiratory illnesses, skin diseases, and zoonotic infections from mycobacteria. Human health is also threatened by the overuse of antibiotics and chemicals in aquaculture, leading to potential antibiotic-resistant pathogens. Chemicals like organophosphates and OPEs in pesticides and flame retardants can accumulate in seafood, which may pose long-term health risks, especially in high seafood-consuming countries. Labour exploitation is another major concern, especially in Southeast Asia, South America, and even the United States. Workers, including trafficked individuals, face hazardous conditions, low wages, and in extreme cases, slavery-like treatment. Reports from Thailand and Chile reveal abuses, while in Colorado, prisoners were exploited for cheap labour in tilapia farming, sparking public outcry and legislative change. What is interesting from an injustice perspective is that there is a clear imbalance in risk levels. The direct and indirect risks from production are heavily weighted on poorer communities, whereas the benefits and risks of fish and prawn consumption appear to be significantly weighted towards wealthier communities. The next section looks at the wider ecological injustices that are associated with the farming of prawns, salmon, and tilapia.

Ecological Justice

Global aquaculture topped "capture fisheries" for the first time in 2022, meaning that more of the world's fish and seafood were produced in fish farms than were wild caught (FAO 2024a, p. 18). The FAO further state that "it is vital that aquaculture development be urgently accelerated in countries which have an aquaculture potential" (2024, p. 18). Ritchie et al. (2024) calculate that the fresh water use, land space needed, and carbon emissions per 100 g of protein for farmed prawns are 2,380 litres, 2 m^2 and 26.09 kg. The equivalent for farmed fish are 1,619 litres, 3.7 m^2 and 7.61 kg.[2] Unsurprisingly, although the water requirements are significantly higher for farmed prawns and fish than they are for farmed beef and pig meat (see Chap. 2), the land-use requirements are significantly lower. As more aquaculture is brought on shore, however, these numbers are likely to change. With the FAO's exhortation to increase aquaculture of this type, the impact of water use, and waste generation will be increased significantly.

Space

The development of aquaculture facilities for species such as salmon and prawns often involves transforming natural coastal or estuarine habitats into controlled environments. These modifications have severe consequences for local ecosystems, as they replace complex habitats with simplified and heavily managed environments that lack the biodiversity and structural features necessary to support native species. In regions like Norway and Chile, salmon farms disrupt fjord ecosystems and reduce habitat quality for endemic marine species, many of which are extremely sensitive to changes in water quality and habitat structure (Naylor et al. 2000). Bøhn et al. (2024) have found that the waste feed which accumulates around salmon farm cages is having an impact on wild cod stocks, both in terms of changing cod behaviour (the fish are drawn to the lights and food of the fish farm) and in increased mortality rates (either directly through ingestion of antibiotics and other drugs given to the salmon, or indirectly if the cod's natural foodstuffs are killed).

Similarly, prawn farms are typically established in coastal mangrove areas, leading to extensive mangrove deforestation and loss of habitat for countless marine and terrestrial species. Su et al. (2024) point out that "Despite their critical ecological roles, these invaluable ecosystems face global decline due to numerous anthropogenic threats, such as coastal development, sewage discharges, and aquaculture" (2024, p. 2). Mangroves serve as breeding and nursery grounds for many fish, birds,

[2] Both "farmed prawn" and "farmed fish" are amalgamated categories, and so there are no data available for king prawns, tiger prawns, salmon, or tilapia.

and invertebrates, and their damage or destruction leads to a cascading effect of biodiversity loss (Primavera 2006). Primavera also points out that many "aquaculture farms in Asia stock wild-caught juveniles rather than hatchery-reared post-larvae derived from wild spawners or brood-stock" (p. 535) and that there are significant issues caused by "bycatch" (i.e., catching fish which are not the target but are nonetheless killed).

Aquaculture farming practices impose heavy environmental burdens on surrounding wildlife, leading to declines in biodiversity and the displacement of local species. Species introduced into marine cages and pens—often in high densities—exert ecological pressures on surrounding organisms through competition, predation, and the alteration of local food webs. Two decades ago, McGinnity et al. (2003) were warning that escaped farmed salmon, which are not native to many of the environments in which they are introduced, can hybridise with wild salmon populations, weakening the genetic fitness of wild stocks and reducing biodiversity (McGinnity et al. 2003). These fears proved to be unfounded after the escape of nearly 50,000 immature farmed salmon after Carradale North Farm was dislodged during a storm in 2020, but the Scottish Government report added the significant caveat that this lack of interbreeding "is likely a result of the farmed fish being immature and therefore unable to breed with wild individuals in the year of the escape" (Gilbey et al. 2022). Other sources (see, for example Einum and Fleming 1997; Sylvester et al. 2018; Wringe et al. 2018) have demonstrated that where adult fish escape, inter-breeding, and hybridisation are commonplace. This not only endangers wild populations by reducing genetic diversity but also disrupts ecological balance, as non-native salmon compete with native species for resources, which can lead to their decline. Forseth et al. (2017) concluded that farmed Atlantic salmon which escaped from Norwegian fish farms was the largest threat to wild Norwegian Atlantic salmon, both in terms of the salmon lice infecting the population and in terms of competition for foodstuff and introgression.

It is not just the introgression (hybridisation caused by interbreeding) caused by the millions of farmed salmon which escape which is causing issues. Though there has been considerably less research into the escape of tilapia, McCrary et al. (2007) looked at an escape from a fish farm in Nicaragua. They concluded that the escaped tilapia were having a significant negative impact on native species of fish. This was partly through competition for food, but also related to the tilapia eating *chara* (a type of algae) which were the favoured nesting material for native fish, and infecting native fish with a virus which causes blindness. The end result is that fish species which were once abundant in Nicaraguan waters are now scarce or absent (McCrary et al. 2007).

Ricardo et al. (2018) have discovered similar issues with farmed shrimp in Mexico escaping and out-performing native species but found no significant evidence of introgression.

Place

As we have seen earlier, farmed fish (whether Salmon, Tilapia, prawns, or other species) are vulnerable to sea lice and other pests, because the density of fish is far greater than it is in the wild. Some fish farmers use chemicals and pesticides to control these pests, and the residue enters the sea, and has an impact on other species. Similarly, much as in urban areas, the population density of farmed fish makes the faecal waste much more problematic than in dispersed populations. This too enters the sea and causes contamination. Furthermore, farmed fish are fed human-processed food which, though free of some of the contaminants found in the wild, is nonetheless processed. There are new developments in commercial fish food which are intending to use waste vegetable and fruit pulp as one of the raw materials, which will potentially help to reduce food waste.

Prawn, or shrimp farming in Southeast Asia (which is the source of the majority of the world's farmed prawns), has been linked to unsustainable practices such as "mangrove destruction, sedimentation, saltwater interruption, loss of biodiversity and contamination" (Shahriar et al. 2019). In addition to this, there are significant environmental impacts involved in transporting the (usually frozen) prawns and shrimp from their production area to the markets.

Removing farmed fish and prawns from the human food supply would necessitate one of two things. Either a reduction in the global demand for prawns or, and much more likely, a dramatic increase in the hunting of wild fish and prawns. This would do little to improve ecological injustices. Wild tiger prawns are often caught using bottom trawl (or otter trawl) nets which the Australian Fisheries Management Authority (AFMA) says can damage sensitive seabed environments such as seagrass and risks unwanted bycatch (other unwanted species which are caught in the trawling process). The AFMA has put in place monitors and guidelines to minimise this environmental damage.

Wild, line-caught salmon incur very little in terms of environmental costs in their production, other than the carbon emissions from fishing vessels, but the practice is not without controversy. Over-fishing is a significant issue in some areas such as Greenland and the Faroe Islands, and the climate crisis is reducing the numbers of salmon available (Linden 2022). Not all wild-caught salmon is line-caught, however, and gillnet fishing which damages many of the fish which pass though the nets is widely used (Kendall et al. 2009). Wild-caught Tilapia is not widely available commercially.

Waste

The final part of this section will look at the impact of different types of waste from fish farms on their local ecosystems. The use of antibiotics in aquaculture, particularly in high-density prawn and salmon farms, introduces large quantities of pharmaceutical compounds into ecosystems. Antibiotics that are not fully absorbed by the farmed animals leach into surrounding waters, leading to the spread of antibiotic-resistant bacteria, which poses ecological risks for aquatic species that are part of the same ecosystem (Cabello et al. 2016). These bacteria can outcompete native bacterial species, disrupting natural microbial communities essential to nutrient cycling and primary productivity.

Even where the antibiotics are completely ingested by the farmed fish and prawns, antibiotic resistance impacts predator–prey dynamics within these ecosystems. Predators that ingest resistant bacteria through prey are at risk of infection by pathogens that they can no longer resist, potentially leading to declines in predator populations and subsequent imbalances in food web structures (Kümmerer 2009). While this is unlikely to occur within the confined of a fish farm, where valuable stocks are protected, we have already seen that countless farmed fish and prawns escape into the wild every year, where they are vulnerable to predation.

In addition to antibiotics entering the natural environment, prawn and salmon farms contribute significantly to nutrient pollution, particularly nitrogen and phosphorus, due to uneaten food, faeces, and chemical inputs. The release of these nutrients contributes to eutrophication, stimulating harmful algal blooms (HABs) that reduce oxygen levels in water, creating anoxic and hypoxic conditions that are lethal to many aquatic organisms (Naylor et al. 2000). This process often results in "dead zones" where aquatic life cannot survive, destroying ecosystems and eliminating habitat for a variety of species.

Not only do HABs reduce oxygen, but they can also produce toxins harmful to fish, marine mammals, and seabirds. Blue-green algae (cyanobacteria) which occur in fresh water and estuaries, produce cyanotoxins which are known for their toxicity to humans and other species (Bownik 2010). Red algae blooms, which occur in seawater, may produce brevetoxins, which have caused death in manatees, dolphins, and some species of fish-eating marine birds (Beasley 2020).

What this section has shown is that in recent years, aquaculture has surpassed wild-capture fisheries, with fish farms now producing more seafood than traditional fishing. The FAO encourages this growth, especially in countries with potential for aquaculture expansion. However, this shift brings significant ecological consequences. Farmed prawns, salmon, and tilapia require substantial water and land resources and in the case of prawns, have high level of carbon emission. Transitioning more farms onshore could alter resource demands but also increase waste issues.

Ecologically, aquaculture disrupts sensitive habitats, for example Norway's fjords and Southeast Asia's mangrove forests, replacing complex natural systems with simplified, heavily managed environments. This reduces biodiversity and affects species dependent on these habitats. The frequent escape of farmed fish also poses risks: non-native salmon and tilapia outcompete wild species, reduce genetic diversity, and spread diseases. Further, antibiotics and nutrient-rich waste from farms drive the growth of harmful algal blooms, creating oxygen-depleted "dead zones" that threaten marine life and destabilise aquatic ecosystems.

The ecological justice impacts of aquaculture are profound and far-reaching. By prioritising economic efficiency, aquaculture has systematically undermined the intrinsic rights of ecosystems and non-human life forms to exist and flourish independently of human exploitation. From habitat destruction to biodiversity loss, chemical contamination, and the spread of invasive species, aquaculture practices disrupt ecosystem integrity and violate principles of ecological justice. Were the trade in farmed fish and prawns to cease to exist, there would be clear ecological benefits. However, as identified earlier, human behaviour is likely to mitigate this otherwise positive impact, by increasing the demand for wild-caught equivalents.

The next section of the chapter will look at the species justice aspects of the trade in farmed fish and prawns, focusing on the specific impact on those species.

Species Justice

The question of sentience looms large in discussions about the ethical aspects of breeding and consuming prawns and fish. Sentience in this context implies a level of cognitive and sensory awareness that should factor into ethical considerations. Many people, for example those adhering to a vegan diet, argue that eating any sentient creature (one capable of feeling pain and suffering) is morally and ethically wrong, and would not eat fish or prawns. Research increasingly supports the sentience of fish and crustaceans, suggesting that they possess the capacity to experience pain, stress, and possibly emotions (Braithwaite 2010).

The trigger for ethical debate is largely about the ways in which prawns and fish are killed before being processed into food, although Petrossian and Clarke (2013) and others have considered the impact of illegal fishing, which is naturally not included in official statistics. They conclude that "illegal fishing heavily impacts not just the species concerned, but also human populations, because it can devastate communities that rely on fishing for employment and on fish for survival" (2013, p. 85). It is also worth noting the impact of corruption here, as illegally wild-caught products may be fraudulently labelled as legal, farmed products.

Neither prawns nor fish are included within the UK's Animal Welfare Act 2006. Indeed, during the Parliamentary debates leading to the Act, there was considerable pressure to exclude fish, both freshwater (coarse and fly), and sea fish. There are extensive legal regulations around the farming of both, but the welfare aspect is not the leading concern. The introduction of the Animal Welfare (Sentience) Act 2022, while being lauded in the press as potentially putting a stop to boiling lobsters alive (see for example Horton 2021),[3] actually devolves the power to decide on the sentience of species to a new Animal Sentience Committee. It will not be mandatory for Government to follow the Committee's reports, however, it just has to demonstrate that is has given "due regard" to the findings of sentience.

In her study of fish cognition, Brown (2015) found that fish demonstrate complex types of behaviour associated with sentience, such as problem-solving and social learning, which suggest an awareness of their surroundings and a capacity for suffering, and that cichlids (the family

[3] Note: The boiling of live lobsters is banned in New Zealand, Norway, and Switzerland (European Parliament, Parliamentary Question E-003646/2021, July 2021).

of fish which includes tilapia) "show lower levels of aggression towards familiar than unfamiliar neighbours" (2015, p. 10)—what Frostman and Sherman (2004) call the "dear enemy" effect).

With fish, Kupsala et al. (2013) agree that "animal welfare as a whole is a contested concept in science and in society" (p. 120) although there is a significant and growing body of study which has demonstrated that fish are indeed sentient creatures (see, for example Kemmerer 2016; Woodruff 2018; Bauer 2019; Mason and Lavery 2022).

Pedrazzani et al. call shrimp (prawn) farming "one of the most controversial agri-food systems" (p. 808), but also point out that while other decapods have been shown to be sentient, "this does not seem to be the case with penaeid crustaceans" (p. 809). Wuertz et al. (2023) make the valid point that "aquaculture growth is outpacing welfare knowledge" (p. 11) and that this is particularly true with decapods.

For prawns and other crustaceans, similar research is emerging. Barr et al. (2008) assessed prawns for nociception, which is the ability to detect noxious substances. This is not enough to demonstrate sentience per se and is common across many species. However, Barr et al. (2008) also showed that in addition to exhibiting nociception, prawns may retain memories of negative experiences, an indication that they too experience suffering, and that their responses suggest "that higher processing is involved in the behavioural output" (2008, p. 750).

Having established that it is more likely than not that both fish and prawns are sentient creatures, we must now turn out attention to ethical challenged around animal welfare and the commodification of sentient creatures.

There is no reason the Five Freedoms outlined in Chap. 2 should not apply equally for farmed land-based animals and farmed water-based animals. Intensive aquaculture conditions often compromise welfare by exposing animals to crowded enclosures, which restricts their mobility in contravention to the Five Freedoms. Farmed fish and prawns are commonly held in high-density systems that limit their natural behaviours, leading to physical and psychological stress (Ashley 2007), and the higher the stocking density, the higher the stress levels (Calabrese et al. 2017).

Different species of fish are killed in different ways, but one similarity is the use of stunning (rendering the fish unconscious) before death. Indeed, for fish farmed within the EU, Regulation 1099/2009 (the Protection of Animals at the Time of Killing Regulations) makes it a legal requirement that "Animals shall be spared any avoidable pain, distress or suffering during their killing and related operations" (Article 3(1)). This is the only part of these regulations which applies to farmed fish, however, and they do not apply at all to prawns.

Fish may be stunned by electrical stunning (being subjected to an electrical current while either in or out of water) or by percussive stunning (being struck on the head by a blunt object either manually or automatically) (Mercogliano et al 2024). While stunning is widespread, it is not without risk. Researchers have discovered that for Atlantic salmon (Robb and Roth 2003) and other farmed fish (van de Vis et al. 2014) an EEG (electroencephalogram, a test measuring brain activity) is not always accurate, and furthermore the effects of electrical stunning are transient, meaning there is an enhanced risk the fish will not be unconscious while it is killed and gutted. van de Vis et al. also said that accurate "stress-physiological and neurophysiological" measurements of fish welfare may not ever be possible in a commercial setting like a fish farm, "therefore, only behavioral and physical measurements can be performed" (2003). Sundall et al. (2024) investigated the effect of electrical and percussive stunning on Tilapia and found that while electrical stunning was immediate, the effects were transient, and so farm operatives needed to work quickly to ensure the fish did not regain consciousness. Mercogliano et al. (2024) similarly note that:

> The stunning application might be ineffective when immobility occurs before consciousness is lost. In this case, there is a risk that exsanguination and evisceration are performed on a conscious animal. If the period of unconsciousness is too short, fish may recover consciousness before being killed (2024, p. 3)

However poor the lot of farmed fish may be, Ungoed-Thomas says that "Wild fish typically die from lack of oxygen on the deck of a trawler after they are caught, are rarely stunned, and may be eviscerated before death" (2022).

There is far less published research on prawn welfare at the time of harvesting, perhaps due to the fact that evidence of decapod sentience lags some way behind that of fish sentience (Elwood 2012). Fu et al. (2014), looked at common ways of killing whiteleg shrimp (the alternate name for King Prawns) to assess the implication on both welfare and the quality of the prawn meat. The authors argue that "good welfare means good production" and thus the better the quality of the prawn meat, the more humane the method of killing. The methods they used were the most common. Both involve, 24 h pre-killing starvation and the prawns were then killed either by individual beheading or ice water immersion. The link between beheading and death need not be explained, and ice water immersion is used because it induces thermal shock (whiteleg shrimp are warm water animals) rendering the animal insensible to pain at the time of death. Fu et al. (2014) "suggested that ice water immersion is a suitable processing method for whiteleg shrimp" as the meat quality, and thus the welfare, were better.

This section of the chapter has looked at some of the key ethical and welfare challenges of fish and prawn farming, focusing on the concept of sentience and its implications for the treatment of these animals. Increasing evidence supports the sentience of both fish and crustaceans, suggesting that they possess the capacity to experience pain, stress, and even complex behaviours such as problem-solving and social learning. This recognition increases the arguments for placing moral obligations on humans to address the welfare of these species within aquaculture practices.

Despite these advancements in understanding fish sentience, welfare practices in aquaculture remain inconsistent and inadequate. While regulatory frameworks such as the EU Protection of Animals at the Time of Killing Regulations address basic aspects of humane killing of farmed fish, these protections are extremely limited and do not extend at all to crustaceans. Additionally, common methods of stunning, while intended to minimise suffering, are often transient in effect, posing a risk of conscious suffering during slaughter. For prawns, the lack of definitive recognition of sentience has led to limited research and weaker welfare standards, perpetuating harmful practices such as beheading and prolonged exposure to ice water immersion.

The treatment of wild-caught fish and prawns may be no better, however, and if Ungoed-Thomas's assertions (2022) are correct, it may in fact be worse. The monitoring of this is considerable harder, however, given the mobile and inaccessible nature of high seas fishing. While the removal of farmed fish and prawns from the human diet, and the ending of the farming industry associated with it, would appear therefore to have an overwhelmingly positive impact on the welfare of fish and prawns, there is therefore the risk that the industry will revert to wild-caught, and increase the level for animal suffering.

One option is for the aquaculture industry to take significant steps towards ensuring that the welfare of fish and prawns is respected as a moral imperative, rather than being subordinated to economic considerations. The next part of the chapter will focus on those economic considerations.

Economic

The aquaculture industry has become a critical component of the global food economy, with farmed fish and prawns contributing significantly to global seafood supply. As we saw earlier in the chapter, the FAO (2024a) has calculated that aquaculture surpassed wild-capture fisheries in 2022, indicating its growing importance in meeting the world's protein demands. Salmon, tilapia, tiger prawns, and king prawns/whiteleg shrimp are among the most economically significant species in aquaculture (FAO 2024b). This section of the chapter examines their economic impacts, focusing on their contributions to national and global economies, trade, employment, and sustainability challenges. The UN FAO's most recent SOFIA report (State of World Fisheries and Aquaculture) says that "an estimated 61.8 million people were employed in the primary sector of fisheries and aquaculture in 2022, down from 62.8 million in 2020" (FAO 2024a).

Salmon farming, particularly Atlantic salmon, is a highly valuable segment of the aquaculture industry. Norway leads global production, accounting for over 50% of farmed salmon exports (FAO 2024a). In the year to December 2023, Norway's salmon exports were over NOK122bn (US$11bnbn), which was almost 75% higher than in the COVID-19-affect year to end of 2020 (NSC 2024). The rise in exports is driven by increased demand in Europe, North America, and Asia (NSC 2024). This income is vital to Norway's coastal communities, where salmon farming provides direct and indirect employment for thousands of workers (MOWI 2023) and supports ancillary industries such as feed production, logistics, and processing.

As we have already, seen, Chile is the second-largest salmon producer in the world (Carrasco-Bahamonde and Casellas 2024), with its industry contributing $6.4 billion to the economy in 2023 (Molinari 2024), when the salmon industry was also the country's second-largest export commodity (Gairn 2024). Similarly to Norway, the salmon industry in Chile directly employees a large number of people and supports an even larger number of ancillary industries. In Scotland too, the salmon farming industry "supports thousands of jobs, particularly in rural and coastal communities, and contributes significantly to Scotland's GDP" (Scottish Salmon 2024).

Tilapia farming is a key source of affordable protein in developing and emerging economies—indeed, El-Sayed (2020) remarks that is it considered as "low-cost, high-quality protein for the poor" (2020, p. 190). China leads global production, accounting for around a quarter of farmed tilapia, followed by Indonesia, Egypt, and Brazil (FAO 2024a), and the industry as a whole was valued by the Dutch multinational food and agribusiness bank Rabobank at around $4 billion in 2023, with significant growth forecast for 2024 and 2025 (Sharma and Nikolik 2023). In Africa, tilapia farming is vital for food security and rural development. Egypt, Africa's largest tilapia producer, produces 1.5 million tons annually (Rossignolia et al. 2023). This contributes around $2 billion to its GDP (FAO 2024a) and employs 150,000 people in full time employment and countless others in part time work (El-Sayed and Fitzsimmons 2023). Similarly, in Uganda and Nigeria, tilapia farming has enabled small-scale farmers to access global markets, improving livelihoods and reducing poverty (FAO 2024a).

King prawns may dominate global shrimp farming due to their high yield and rapid growth, but tiger prawns too are a major aquaculture species in Asia, particularly in India, Thailand, and Vietnam. According to the FAO's FishStatJ database, over six million tons of king prawns and five million tons of tiger prawns are farmed every year (FAO 2024c) mostly in South and Southeast Asia. There are also significant South American producers. Ecuador, for example exports close to US$6bn/1.3 mt of king prawns a year, according to ProEcuador, the government export and import body (ProEcuador 2023). In India, the Ministry of Commerce and Industry announced in the summer of 2024 that "Frozen shrimp, which earned ₹40,013.54 crore (US$4881.27million), retained its position as the top item in the seafood export basket, accounting for a share of 40.19% in quantity and 66.12% of the total dollar earnings" (PIB Delhi 2024). This is particularly important given the significant role of women in prawn cultivation in India, although a 2024 NGO investigation found examples of debt bondage, worker exploitation, dangerous working conditions, and significant environmental damage in India's prawn industry (CAL 2024).

There are then significant economic interests enmeshed in the fish and prawn farming industry. With any intensive industrial fish farming operation, however, the risks of pests, escapes, and disease are far higher than for the equivalent wild-caught industry. Disease outbreaks, such as sea lice infestations, significantly impact productivity and profitability. Costello (2009) estimated that in salmon farming countries affected by sea lice, the cost of control is around 6% of the market size, running to nearly $500 million annually (Costello 2009).

In prawn farming, a disease called Early Mortality Syndrome (EMS) which is also known as acute hepatopancreatic necrosis disease (AHPND), was discovered in 2009. EMS/AHPND has a high mortality rate and appears to be spread from hatcheries and nurseries to farms (Poolkhet et al. 2023). EMS/AHPND has since cost the industry in excess of US$10bn (based on Zorriehzahra and Banaederakhshan's (2015) estimate of US$1bn a year). A side effect of the EMS/AHPND outbreak is the unfortunate increase in the use of antibacterial and antimicrobial substances in some farms. In addition to the specific threat posed by EMS/AHPND, Compassion in World Farming conducted a meta-study of prawn mortality rates for different stocking densities and found that all eleven of the studies they looked at showed a marked increase in shrimp mortality with increased stocking density (CIWF 2024, pp. 6–7). Decreasing farmed shrimp stocking density would therefore increase welfare, and income—because, as Fu et al. noted "good welfare means good production" (2014).

Farmed prawns and fish undoubtedly have a much greater environmental impact than their wild-caught equivalents, but they are a much more reliable source of income for the producer and are considerably cheaper for the consumer. Fresh, wild-caught salmon fillets, for example, as sold in UK supermarkets are significantly higher in price than the cost of the farmed equivalent and, where they are available, wild-caught prawns are also significantly higher in price (see Table 4.1).

What Table 4.1 demonstrates is that, for UK consumers at least, the average supermarket price of fresh farmed salmon fillets is £20.50 per kilogramme, and fresh wild salmon fillets are £38.30 per kilogramme, almost ninety per cent higher. Farmed king prawns averaged at £22.00 a kilogramme, however at the time of the sampling no wild-caught alternative was available in any major UK supermarket, but online only retailers sell frozen (rather than chilled) wild-caught prawns for around £35.00 per kilogramme, a sixty per cent increase. What this suggests is that the

Table 4.1 Comparative costs of wild-caught and farmed salmon and king prawns in UK supermarkets

	Chilled raw salmon		*Chilled raw king prawns*	
Supermarket	*Wild*	*Farmed*	*Wild*	*Farmed*
Aldi	N/A	18.71/kg	N/A	15.93/kg
Asda	38.64/kg	18.75/kg	N/A	21.25/kg
M & S	36.40/kg	24.00/kg	N/A	24.20/kg
Morrisons	N/A	20.45/kg	N/A	21.25/kg
Sainsbury	36.96/kg	19.79/kg	N/A	21.21/kg
Tesco	38.70/kg	19.04/kg	N/A	21.18/kg
Waitrose	40.68/kg	22.89/kg	N/A	28.79/kg

Note Data was taken from supermarket websites in November 2024, and excludes any promotions. Lidl (www.lidl.co.uk) was not included as their website does not include prices

immediate removal of farmed salmon and prawns from the market would please a significant financial burden on consumers. As consumer awareness of environmental and ethical issues grows, markets are demanding certifications such as the Aquaculture Stewardship Council (ASC) label, which requires adherence to environmental and social standards. While certification improves market access, it also increases production costs, potentially pricing small-scale farmers out of global markets (Bush et al. 2013).

This section has focused on the economic impact of specific aspects of aquaculture, focusing on salmon, tilapia, tiger prawns, and king prawns/whiteleg shrimp. With aquaculture having surpassed wild-capture fisheries as the dominant source of seafood in 2022 (FAO 2024a), these industries play a critical role in meeting global protein demands and supporting economies worldwide. A sudden halt would have profound implications, likely resulting in significant economic disruption. Farmed salmon is a cornerstone of national economies like Norway and Chile, generating over $11 billion and $6.4 billion annually, respectively, and supporting thousands of jobs in rural and coastal communities. Scotland also relies heavily on the industry, where it sustains ancillary sectors like feed production, logistics, and processing. The removal of farmed salmon would destabilise these economies, particularly in coastal regions, while skyrocketing prices for wild salmon—already 90% higher than farmed salmon—would burden consumers.

Tilapia, a key source of affordable protein in developing countries, contributes $2 billion to Egypt's GDP and supports rural development in Africa and Asia. Halting tilapia farming would exacerbate poverty in regions reliant on small-scale aquaculture, worsen food insecurity, and force higher reliance on more expensive proteins. Similarly, ceasing tiger prawn and whiteleg shrimp farming, which generate billions annually in exports from countries like Ecuador and India, would devastate livelihoods, particularly in rural Asia, South America, and coastal areas where women play a significant role in the industry.

Additionally, replacing farmed seafood with wild-caught alternatives is neither sustainable nor practical. Wild stocks cannot meet current demand without risking overfishing, while wild-caught products are significantly more expensive. For instance, wild prawns cost approximately 60% more than farmed equivalents. The immediate removal of farmed products would impose substantial financial burdens on consumers and further strain global seafood supplies.

Although the aquaculture industry has significant environmental and ethical challenges, including disease outbreaks, poor animal welfare, and ecosystem damage, a sudden halt would economically harm millions of workers and communities while making seafood unaffordable for many consumers. Thus, from a purely economic perspective, a complete cessation of this industry would cause widespread economic harm and is not a solution unless viable alternative income streams can be developed.

Future

If the global aquaculture industries for salmon, tilapia, king prawns, and tiger prawns were to cease, the ripple effects would profoundly impact global food systems, economies, and ecosystems. These industries supply a significant proportion of the world's seafood, generate billions of dollars in revenue, and sustain millions of jobs. The loss of these sectors would require the development of alternative protein sources, income streams, and strategies to mitigate social and economic disruptions. This section examines the challenges and potential pathways forward, considering alternate protein sources, environmental impacts, and economic adaptation.

The aquaculture sector in general, and in particular the aspects covered in this chapter, is currently of significant importance in meeting global protein demands. As of 2022, aquaculture accounted for over 50%

of seafood consumed globally, surpassing wild-capture fisheries (FAO 2024a). These species alone provide affordable, nutrient-dense protein to billions of people, particularly in low- and middle-income countries, and particularly El-Sayed's "low-cost, high-quality protein for the poor", tilapia (El-Sayed 2020).

Without these industries, there would be a significant protein deficit, particularly in regions reliant on aquaculture for food security. The FAO (2024a) estimates that fish and seafood provide 17% of global animal protein consumption, with higher dependency in Asia and Africa. The cessation of these industries without alternative options would exacerbate malnutrition, increase food insecurity, and put immense pressure on alternative protein sources, which may not be readily available or affordable.

There are four broad categories of alternate sources of protein—land-based animals, sea-based animals, plants, and cultured proteins. For the reasons which were explored in Chap. 2, replacing farmed fish and seafood with increased farming of animals would cause as many, if not more, problems than it would solve. Land-based farming of animals is unsustainable, particularly if we consider all three strands of sustainable development, rather than just the economic side. If land-based animal farming is not a viable alternative to water-based animal farming, the next alternative to consider is increasing wild fishing to fill the gap caused by the lack of farmed fish.

Unfortunately, wild fishing also has significant and inherent problems. In 1974, the FAO estimated that 90% of global fishery stocks were within biologically sustainable levels (FAO 2022). By 2017 this has dropped to 65.8%, and by 2019 to 64.6% (FAO 2022). By 2024, this has dropped further to 62.3% (FAO 2024a). Despite the FAO's acknowledgement that aquaculture has out-produced "capture fisheries" since 2022 (FAO 2024a, p. 18), the percentage of sustainable fish stocks continues to decrease. Any move to expand wild-capture fisheries to replace farmed seafood would therefore accelerate biodiversity loss, disrupt marine ecosystems, and exacerbate the ecological crisis.

The third possible protein alternative to farmed fish is plant-based protein, such as that which can be obtained from legumes (soy, lentils, peas, beans). Geographically, the areas which lead the world in fish and prawn farming (broadly South and Southeast Asia, South America and increasingly north Africa and some parts of Europe) are not yet significant players in aspects of the global plant-based protein market. For soya, 80%

is currently produced in Brazil, the United States and Argentina (USDA 2024), for lentils, Canada, India, and Australia produce 69% (Kaale et al. 2023) and for chickpeas (also called garbanzo beans) India alone produced three quarters of the global crop (Çelik 2020). Economically, therefore, this would not serve as a direct replacement for the aquaculture industry, but more of a diversion. For consumers who are not yet prepared to move from fish to vegetable protein, there are already a range of ultra-processed foods (UPF) on the market which mimic the taste and texture of fish and prawns but are made of vegetable protein. The issues with UPFs were discussed in Chap. 2, and it is widely accepted that consumption of such foods should not be increased (see, inter alia, Avesani et al. 2023; Gangopadhyay 2023; Nilson et al. 2023; Samuthpongtorn et al. 2023).

The final alternative protein source is cultured protein, in the form of lab-grown seafood. As of the end of 2024, there are a number of companies across the world working on the production of this type of food, where cells from a fish are cultured in a laboratory. This industry is still in its infancy, however, and even assuming high levels of consumer acceptance (see Kantono et al. 2022), it will be some time until it is scaled up to the extent where it will match the current demand for farmed fish and seafood. An even more esoteric future development may be the 3D printing of fish and seafood. Gurel et al. (2024) explored the possibilities of this technology, and concluded that although it has potential, "it is essential for researchers, policymakers, industry and ethicists to work collaboratively to ensure clear ethical guidelines and regulations that promote the responsible and sustainable development and the use of bioprinted muscle tissue" (2023, p. 12).

Regardless of these alternate sources of protein, a significant area of concern is going to be the immediate loss of income for those millions of people who reply on aquaculture. Diversifying income streams would be essential. Coastal tourism, renewable energy projects (e.g., offshore wind farms), and seaweed farming are potential avenues for economic diversification. Seaweed, in particular, has a growing market for use in food, biofuels, and bioplastics and requires minimal inputs compared to aquaculture (Kim et al. 2015). The speed at which kelp grows would make it a short-term cash crop, and a viable replacement for fish, and the World Bank has predicted that the global seaweed market will be worth close to US$12bn by 2030 (World Bank 2023).

The cessation of salmon, tilapia, king prawn, and tiger prawn farming would have profound short- and medium-term implications for global food security, economies, and ecosystems. While there are clear environmental benefits, including reduced coastal degradation and pollution, the immediate consequences—protein deficits, economic disruption, and increased reliance on less sustainable alternatives—would outweigh the advantages. Transitioning to plant-based proteins, lab-grown seafood, and diversified livelihoods offers long-term solutions, but the nascent nature of many of these new technologies, and what appears to be a lag in human acceptance of them means that these require significant investment, technological advancement, and global cooperation before they become mainstream. Balancing environmental, ethical, and economic considerations will be critical in shaping a sustainable future for global food systems.

Conclusion

The chapter has explored the issues associated with the farming of tiger prawns, king prawns, Atlantic salmon, and tilapia. These species, central to global seafood supply, highlight the inherent tensions between economic growth, environmental sustainability, and social equity. The analysis reveals that while aquaculture offers solutions to pressing global challenges such as food security and poverty alleviation, it simultaneously exacerbates ecological degradation, labour exploitation, and ethical concerns regarding animal welfare.

A recurring theme throughout the chapter is the disproportionate impact of aquaculture's negative externalities on marginalised communities and vulnerable ecosystems. In Southeast Asia and Latin America, the destruction of mangroves for shrimp farming has displaced local communities and disrupted ecosystems that provide critical services such as carbon sequestration and coastal protection. Similarly, in countries like Norway and Chile, salmon farming has led to pollution, habitat degradation, and the spread of diseases such as sea lice, which threaten both farmed and wild fish populations. These ecological injustices highlight the need for a more equitable and sustainable approach to aquaculture.

Social justice issues are equally prominent. The industry's reliance on low-wage labour, often sourced from vulnerable populations, underscores the ethical challenges of aquaculture. Reports of human trafficking, modern slavery, and exploitative working conditions in the prawn and fish farming sectors reveal a darker side to the industry's success. These injustices are compounded by inadequate legal protections for workers and the prioritisation of profits over human dignity. The chapter underscores the urgent need for stronger regulatory frameworks and enforcement mechanisms to address labour exploitation and ensure fair treatment for all workers in the industry.

From an ecological perspective, the environmental costs of aquaculture are significant. The nutrient-rich waste and chemical run-off from fish and prawn farms contribute to eutrophication, harmful algal blooms, and the creation of dead zones in aquatic ecosystems. Antibiotic overuse not only harms marine life but also poses a public health risk by accelerating the spread of antibiotic-resistant bacteria. The chapter also highlights the ethical implications of farming sentient creatures like fish and prawns in high-density, stressful conditions. The commodification of these animals raises fundamental questions about humanity's responsibility to ensure the welfare of all sentient beings.

Economically, aquaculture plays a vital role in supporting livelihoods and generating revenue. Countries like Norway, Chile, and India depend on the industry for employment and exports, demonstrating its importance to national and global economies. However, the economic benefits of aquaculture are often unevenly distributed. Small-scale farmers and workers in low-income countries frequently bear the brunt of environmental degradation and market volatility, while consumers in high-income countries enjoy affordable seafood. This imbalance underscores the need for policies that promote more equitable economic outcomes.

Looking ahead, the future of aquaculture depends on addressing its environmental, social, and ethical challenges while harnessing its potential for positive impact. Technological innovations, such as closed-loop systems and alternative protein sources like lab-grown seafood, offer promising pathways for reducing the industry's ecological footprint. However, these solutions require substantial investment, global cooperation, and consumer acceptance. Regulatory frameworks must also evolve to address the welfare of farmed species, incorporating the latest scientific findings on sentience and ethical farming practices.

The chapter concludes that from the standpoint of environmental justice, ecological justice, and ecological justice, the cessation of prawn and fish farming has significantly more positive than negative impacts. The one area where the impacts favour some kind of continuation is economically. However, even in this area we do not need to merely accept the status quo ante for, as Fu et al. noted "good welfare means good production" (2014). If we are to continue to farm fish and seafood, a more sustainable approach to aquaculture is needed—one that prioritises ecological restoration, social equity, and ethical treatment of animals. By embracing innovative technologies, strengthening regulations, and fostering global collaboration, the aquaculture industry can evolve to meet the demands of a growing population while minimising its negative impacts. Balancing these priorities will be essential to creating a more just and sustainable future for aquaculture and the planet as a whole.

Detailed Data for Table 4.1

Aldi

Farmed Salmon
Specially Selected Scottish Salmon Fillets
240 g
18.71/kg
Farmed Prawns
The Fishmonger Cooked & Peeled King Prawns (uncooked king prawns N/A)
150 g
15.93/kg

Asda

Wild Salmon
Leap Wild Sockeye Salmon Fillets
220 g
38.64/kg
Farmed Salmon
ASDA Flavoursome 2 Salmon Fillets
240 g
18.75/kg
Farmed Prawns
ASDA Succulent Raw & Peeled King Prawns
160 g
21.25/kg

M & S

Wild Salmon
Wild Alaskan Sockeye Salmon Fillets
220 g
36.40/kg
Farmed Salmon
2 Scottish skin on and boneless salmon fillets
240 g
24.00/kg
Farmed Prawns
Honduran King Prawns
165 g
24.20/kg

Morrisons

Farmed Salmon
Morrisons 2 Salmon Fillets
220 g
20.45/kg
Farmed Prawns
Morrisons Raw Whole King Prawns
200 g
21.25/kg

Sainsbury

Wild Salmon
Wild Fish Discovery MSC Sockeye Salmon ×2
230 g.
36.96/kg
Farmed Salmon
Sainsbury's Skin on ASC Scottish Salmon Fillets ×2
240 g
19.79/kg
Farmed Prawns
Sainsbury's Raw King Prawns ASC
165 g

21.21/kg

Tesco

Wild Salmon
Tesco Finest 2 Wild Caught Sockeye Salmon Fillets
230 g
38.70/kg
Farmed Salmon
Tesco 2 Boneless Salmon Fillets
260 g
19.04/kg
Farmed Prawns
Tesco Raw King Prawns
170 g
21.18/kg

Waitrose

Wild Salmon
No. 1 2 Wild Alaskan Sockeye Salmon Fillets
220 g
40.68/kg
Farmed Salmon
Waitrose 2 Scottish Salmon Fillets
260 g
22.89/kg
Farmed Prawns
Waitrose ASC Extra Large Raw King Prawns
165 g
28.79/kg

Sources

Ashley, P.J. 2007. Fish welfare: Current issues in aquaculture. *Applied Animal Behaviour Science* 104 (3–4): 199–235. https://doi.org/10.1016/j.applanim.2006.09.001.

Avesani, C.M., L. Cuppari, F.B. Nerbass, B. Lindholm, and P. Stenvinkel. 2023. Ultraprocessed foods and chronic kidney disease—double trouble. *Clinical Kidney Journal* 16 (11). https://doi.org/10.1093/ckj/sfad103.

Barr, S., P. Laming, J.T.A. Dick, and R.W. Elwood. 2008. Nociception or pain in a decapod crustacean? *Animal Behaviour* 75 (3): 745–751. https://doi.org/10.1016/j.anbehav.2007.07.004.

Bauer, H. 2019. Fishes: The forgotten sentient beings. *Derecho Animal* 10 (2): 72. https://doi.org/10.5565/rev/da.427.

Beasley, V.R. 2020. Harmful Algal Blooms (Phycotoxins), Elsevier reference collection in earth systems and environmental sciences. https://doi.org/10.1016/B978-0-12-409548-9.11275-8.

Bøhn, T., J.F. Strøm, P. Sanchez-Jerez, et al. 2024. Ecological interactions between farmed Atlantic salmon and wild Atlantic cod populations in Norway: A review of risk sources and knowledge gaps. *Reviews in Aquaculture* 16 (3): 1333–1350. https://doi.org/10.1111/raq.12899.

Bownik, A. 2010. Harmful algae: Effects of alkaloid cyanotoxins on animal and human health. *Toxin Reviews* 29: 99–114. https://doi.org/10.3109/15569543.2010.516464.

Braithwaite, V. 2010. *Do fish feel pain?* Oxford University Press.

Brown, C. 2015. Fish intelligence, sentience, and ethics. *Animal Cognition* 18 (1): 1–17. https://doi.org/10.1007/s10071-014-0761-0.

Burridge, L., J.S. Weis, F. Cabello, J. Pizarro, and K. Bostick. 2010. Chemical use in salmon aquaculture: A review of current practices and possible environmental effects. *Aquaculture* 306 (1–4): 7–23.

Bush, S.R., B. Belton, D. Hall, P. Vandergeest, F.J. Murray, S. Ponte, P. Oosterveer, M.S. Islam, A.P.J. Mol, M. Hatanaka, F. Kruijssen, T.T.T. Ha, D.C. Little, and R. Kusumawatim. 2013. Certify sustainable aquaculture? Certification's limited contribution to sustainable aquaculture should complement public and private governance. *Science* 341 (6150): 1067–1068. https://doi.org/10.1126/science.1237314.

Cabello, F.C. 2006. Heavy use of prophylactic antibiotics in aquaculture: A growing problem for human and animal health and for the environment. *Environment Microbiology* 8 (7): 1137–1144. https://doi.org/10.1111/j.1462-2920.2006.01054.x

Cabello, F.C., H.P. Godfrey, A.H. Buschmann, and H.J. Dölz. 2016. Aquaculture as yet another environmental challenge for developing countries. *The Lancet Infectious Diseases* 16 (7): e127–e133.

CAL. 2024. Hidden harvest: Human rights and environmental abuses in India's Shrimp Industry. Corporate Accountability Lab. https://corpaccountabilitylab.org/hidden-harvest.

Calabrese, S., T.O. Nilsen, J. Kolarevic, L.O.E. Ebbesson, C. Pedrosa, S. Fivelstad, C. Hosfield, S.O. Stefansson, B.F. Terjesen, H. Takle, C.I.M. Martins,

H. Sveier, F. Mathisen, A.K. Imsland, and S.O. Handeland. 2017. Stocking density limits for post-smolt Atlantic salmon (Salmo salar L.) with emphasis on production performance and welfare. *Aquaculture* 468 (Part 1): 363–370. https://doi.org/10.1016/j.aquaculture.2016.10.041.

Carrasco-Bahamonde, D., and Casellas, A. 2024. Evolving blue development discourses and policies: Salmon farming industry and regional making in Chile. *Marine Policy* 163: 106111.

CCI. 2024. Correctional Industries Mission, Colorado Correctional Industries. https://cci.colorado.gov/about.

Çelik, S. 2020. Modelling and estimation of chickpea production in Turkey using artificial neural networks and time series analysis. *International Journal of Engineering and Science* 10 (11): 1–7.

Chang, H.-Y., W.-C. Yang, Y.-J. Xue, M.-Y. Tsai, J.-H. Wang, and G.-R. Chang. 2019. Phthalates and organophosphorus insecticide residues in shrimp determined by liquid/gas chromatography-Tandem mass spectrometry and a health risk assessment. *Marine Pollution Bulletin* 144: 140–145.

CIWF. 2024. Food business: Whiteleg Shrimp corporate ask, compassion in world farming. https://www.compassioninfoodbusiness.com/media/7457892/whiteleg-shrimp-corporate-ask-review-2024.pdf.

Costello, M.J. 2009. The global economic cost of sea lice to salmonid farming. *Journal of Fish Diseases* 32 (1): 115–118. https://doi.org/10.1111/j.1365-2761.2008.01011.x.

Ecoceanos. 2023. Nueva muerte de buzo en labores de la industria chilena exportadora de salmón durante el 2023, 16 October. https://www.ecoceanos.cl/2023/10/nueva-muerte-de-buzo-en-labores-de-la-industria-chilena-exportadora-de-salmon-durante-el-2023/.

Einum, S., and I.A. Fleming. 1997. Genetic divergence and interactions in the wild among native, farmed and hybrid Atlantic salmon. *Journal of Fish Biology* 50: 634–51. https://doi.org/10.1111/j.1095-8649.1997.tb01955.x.

El-Sayed, A.-F.M. 2020. The role of tilapia culture in rural development. In *Tilapia culture*, ed. A-F.M. El-Sayed, 2nd ed., 275–295. Elsevier/Academic Press.

El-Sayed, A-F.M., and K. Fitzsimmons. 2023. From Africa to the world—The journey of Nile tilapia. *Reviews in Aquaculture* 15 (51). https://doi.org/10.1111/raq.12738.

Elwood, R. 2012. Evidence for pain in decapod crustaceans. *Animal Welfare* 21 (S2): 23–27. https://doi.org/10.7120/096272812X13353700593365.

FAO. 2022. The state of world fisheries and aquaculture 2022. FAO: Rome. https://www.fao.org/3/cc0461en/online/cc0461en.html.

FAO. 2024a. The state of world fisheries and aquaculture 2024—Blue transformation in action. Rome. https://doi.org/10.4060/cd0683en.

FAO. 2024b. Fishery and aquaculture statistics—Yearbook 2021, FAO Yearbook of Fishery and Aquaculture Statistics. Rome. https://doi.org/10.4060/cc9523en.

FAO. 2024c. FishStatJ—Software for fishery and aquaculture statistical time series: Global production by production source 1950–2022. www.fao.org/fishery/en/statistics/software/fishstatj.

Forseth, T., B.T. Barlaup, B. Finstad, P. Fiske, H. Gjøsæter, M. Falkegård, A. Hindar, T.A. Mo, A.H. Rikardsen, E.B. Thorstad, L.A. Vøllestad, and V. Wennevik. 2017. The major threats to Atlantic salmon in Norway. *ICES Journal of Marine Science* 74 (6): 1496–1513. https://doi.org/10.1093/icesjms/fsx020.

Frostman, P., and P.T. Sherman. 2004. Behavioral response to familiar and unfamiliar neighbors in a territorial cichlid, Neolamprologus pulcher. *Ichthyological Research* 51: 283–285. https://doi.org/10.1007/s10228-004-0223-9.

Fu, L.-L., X. Chen, and Y. Wang. 2014. Quality evaluation of farmed whiteleg shrimp, *Litopenaeus vannamei*, treated with different slaughter processing by infrared spectroscopy. *Food Chemistry* 151 (15): 306–310. https://doi.org/10.1016/j.foodchem.2013.11.012.

Gairn, L. 2024. Chile: salmon industry was second-largest exporter in 2023, Weareaquaculture, 9 January. https://weareaquaculture.com/news/aquaculture/chile-salmon-industry-was-second-largest-exporter-in-2023.

Gangopadhyay, A. 2023. Ultraprocessed foods and cancer in low-middle-income countries. *Nutrition and Cancer* 75 (10): 1863–1873. https://doi.org/10.1080/01635581.2023.2272814.

Gilbey, J., J. Sampayo, E. Cauwelier, L. Chapman, Y.M. Bradley, and D.J. Morris. 2022. Examination of levels of farm/wild hybridisation in south-west Scotland and north-west England following a large-scale farm salmon escape event in 2020. *Scottish Marine and Freshwater Science* 13 (2). Scottish Government/Riaghaltas na h-Alba. https://doi.org/10.7489/12429-1.

Greenpeace. 2021. CIDH recibe a comunidades de tres pueblos originarios y Greenpeace por expansión de salmonicultura, Fundación Greenpeace Pacífico Sur. https://www.greenpeace.org/chile/noticia/greenpeace/cidh-recibe-a-comunidades-de-tres-pueblos-originarios-y-greenpeace-por-expansion-de-salmonicultura/ (in Spanish).

Gurel, M., N. Rathod, L.Y. Cabrera, S. Voyton, M. Yeo, F. Ozogul, and I.T. Ozbolat. 2024. A narrative review: 3D bioprinting of cultured muscle meat and seafood products and its potential for the food industry. *Trends in Food Science & Technology* 152: 104670. https://doi.org/10.1016/j.tifs.2024.104670.

Heal, R.D., N.A. Hasan, and M.M. Haque. 2021. Increasing disease burden and use of drugs and chemicals in Bangladesh shrimp aquaculture: A potential menace to human health. *Marine Pollution Bulletin* 172: 112796. https://doi.org/10.1016/j.marpolbul.2021.112796.

Holmström, K., S. Gräslund, A. Wahlström, S. Poungshompoo, B.-E. Bengtsson, and N. Kautsky. 2003. Antibiotic use in shrimp farming and implications for environmental impacts and human health. *International Journal of Food Science & Technology* 38: 255–266. https://doi.org/10.1046/j.1365-2621.2003.00671.x.

Horton, H. 2021. Boiling of live lobsters could be banned in UK under proposed legislation. *The Guardian*, 19 November 2021. https://www.theguardian.com/world/2021/nov/19/boiling-of-live-lobsters-could-be-banned-in-uk-under-proposed-legislation.

Iber, B.T., and N.A. Kasan. 2021. Recent advances in Shrimp aquaculture wastewater management. *Heliyon* 7: e08283.

Kaale, L.D., M. Siddiq, and S. Hooper. 2023. Lentil (Lens culinaris Medik) as nutrient-rich and versatile food legume: A review. *Legume Science* 5 (2): e169. https://doi.org/10.1002/leg3.169.

Kantono, K., M. Hamid, M. Malavalli, Y. Liu, T. Liu, and A. Seyfoddin. 2022. Consumer acceptance and production of in vitro meat: A review. *Sustainability* 14: 4910. https://doi.org/10.3390/su14094910.

Kemmerer, L. 2016. Ethics and eating fishes. *Journal of International Wildlife Law and Policy* 19(3). https://doi.org/10.1080/13880292.2016.1204879.

Kendall, N.W., J.J. Hard, and T.P. Quinn. 2009. Quantifying six decades of fishery selection for size and age at maturity in sockeye salmon. *Evolutionary Applications* 2 (4): 523–536. https://doi.org/10.1111/j.1752-4571.2009.00086.x.

Kim, J.K., G.P. Kraemer, and C. Yarish. 2015. Use of sugar kelp aquaculture in long island sound and the Bronx River Estuary for nutrient extraction. *Marine Ecology Progress Series* 531: 155–166. https://doi.org/10.3354/meps11331.

Kümmerer, K. 2009. Antibiotics in the aquatic environment–a review–part I. *Chemosphere* 75 (4): 417–434.

Kupsala, S., P. Jokinen, and M. Vinnari. 2013. Who cares about farmed fish? Citizen perceptions of the welfare and the mental abilities of fish. *Journal of Agricultural and Environmental Ethics* 26 (1): 119–135. https://doi.org/10.1007/s10806-011-9369-4.

LCS. 2017. Capital construction fact sheet: Colorado correctional industries. https://leg.colorado.gov/sites/default/files/images/17cdc_fact_sheet_colorado_correctional_industries.pdf.

Linden, E. 2022. *Fire and flood: A people's history of climate change, from 1979 to the present*. London: Allen Lane.

Mason, G.J., and J.M. Lavery. 2022. What is it like to be a bass? Red herrings, fish pain and the study of animal sentience. *Frontiers in Veterinary Science* 9: 788289–788289. https://doi.org/10.3389/fvets.2022.788289.

McCrary, J.K., B.R. Murphy, J.R. Stauffer Jr., and S.S. Hendrix. 2007. Tilapia (Teleostei: Cichlidae) status in Nicaraguan natural waters. *Environment Biology Fish* 78: 107–114. https://doi.org/10.1007/s10641-006-9080-x.

McGinnity, P., P. Prodöhl, A. Ferguson, R. Hynes, N.O. Maoiléidigh, N. Baker, D. Cotter, B. O'Hea, D. Cooke, G. Rogan, J. Taggart, and T. Cross. 2003. Fitness reduction and potential extinction of wild populations of Atlantic salmon, Salmo salar, as a result of interactions with escaped farm salmon. *Proceedings of the Royal Society of London. Series B: Biological Sciences* 270 (1532): 2443–2450.

Mercogliano, R., A. Avolio, F. Castiello, and M.C. Ferrante. 2024. Development of welfare protocols at slaughter in farmed fish. *Animals* 14 (18): 2730. https://doi.org/10.3390/ani14182730.

Molinari, C. 2024, 15 October. Chile Salmon Council pushes back against overproduction discourse, SeaFoodSource. https://www.seafoodsource.com/news/aquaculture/-you-have-to-put-it-in-perspective-chile-salmon-council-pushes-back-against-overproduction-discourse.

MOWI. 2023. Salmon farming industry handbook 2023. Bergen: MOWI.

Naylor, R.L., R.J. Goldburg, J.H. Primavera, N. Kautsky, M.C.M. Beveridge, J. Clay, C. Folke, J. Lubchenco, H. Mooney, and M. Troell. 2000. Effect of aquaculture on world fish supplies. *Nature* 405: 1017–1024.

Nilson, E.A.F., G. Ferrari, M.L.C. Louzada, R.B. Levy, C.A. Monteiro, and L.F.M. Rezende. 2023. Premature deaths attributable to the consumption of ultraprocessed foods in Brazil. *American Journal of Preventive Medicine* 64 (1): 129–136. https://doi.org/10.1016/j.amepre.2022.08.013.

NSC. 2024. Yearly exports from Norway: Salmon/All produces/All countries, Norwegian Seafood Council. https://en.seafood.no/market-insight/norwegian-trade/year/.

OHCHR. 2019. Cheap seafood endangers fish workers right to food—UN expert, Press Release: United Nations Office of the High Commissioner for Human Rights. https://www.ohchr.org/en/press-releases/2019/02/cheap-seafood-endangers-fish-workers-right-food-un-expert.

Ortiz-Cañavate, B.K., J. Wolinska, and R. Agha. 2019. Fungicides at environmentally relevant concentrations can promote the proliferation of toxic bloom-forming cyanobacteria by inhibiting natural fungal parasite epidemics. *Chemosphere* 229: 18–21.

Petrossian G.A., and R.V. Clarke. 2013. Explaining and controlling illegal commercial fishing: An application of the CRAVED theft model. *The British Journal of Criminology* 54 (1): 73–90. https://doi.org/10.1093/bjc/azt061.

PIB Delhi. 2024. Ministry of Commerce & Industry Press Release 2026456: India's seafood exports touch all-time high by volume in FY 2023–24,

19 June 2024. https://pib.gov.in/PressReleaseIframePage.aspx?PRID=2026456.

Poolkhet, C., V. Boonyawiwat, S. Yaemkasem, and J. Kasornchandra. 2023. Post-larvae movement network of marine shrimp during the 2013 outbreak of acute hepatopancreatic necrosis disease in Thailand. *Preventive Veterinary Medicine* 210: 105796. https://doi.org/10.1016/j.prevetmed.2022.105796.

Primavera, J.H. 2006. Overcoming the impacts of aquaculture on the coastal zone. *Ocean & Coastal Management* 49 (9–10): 531–545.

ProEcuador. 2023. Ecuadorian Shrimp Industry Report. https://www.proecuador.gob.ec/.

Ricardo, P-E., J.A. Medina-Espinoza, A. Max-Aguilar, and C.J. Saucedo-Barrón. 2018. Genetic tracing of farmed shrimp (Decapoda: Penaeidae) in wild populations from a main aquaculture region in Mexico. *Revista de Biología Tropical* 66 (1).

Ritchie, H., P. Rosado, and M. Roser. 2024. Environmental impacts of food production, our world in data. https://ourworldindata.org/environmental-impacts-of-food.

Robb, D.H.F., and B. Roth. 2003. Brain activity of Atlantic salmon (Salmo salar) following electrical stunning using various field strengths and pulse durations. *Aquaculture* 216 (1–4): 363–369. https://doi.org/10.1016/S0044-8486(02)00494-5.

Romero, G. 2022. The ugly truth about the Tilapia Boom. https://www.delish.com/food-news/a40654096/tilapia-boom-ugly-truth/.

Rossignolia, C.M., T. Manyisea, K.M. Shikukua, A.M. Nasr-Allahb, E.B. Dompreha, P.J.G. Henrikssona, R.D. Lama, D.L. Lazoa, N. Trand, A. Roeme, A. Badrf, A.S. Sbaayb, R. Moruzzog, R. Tilleya, H. Charo-Karisab, and A. Gasparatosc. 2023. Tilapia aquaculture systems in Egypt: Characteristics, sustainability outcomes and entry points for sustainable aquatic food systems. *Aquaculture* 577: 739952. https://doi.org/10.1016/j.aquaculture.2023.739952.

Samuthpongtorn, C., L.H. Nguyen, O.I. Okereke, D.D. Wang, M. Song, A.T. Chan, and R.S. Mehta. 2023. Consumption of ultraprocessed food and risk of depression. JAMA Network Open 6 (9): e2334770. https://doi.org/10.1001/jamanetworkopen.2023.34770.

Scottish Salmon. 2024. Culture. https://www.scottishsalmon.co.uk/culture/.

Shahriar, A., B. Dhrubo, and H. Sazzad. 2019. Environmental impacts of commercial shrimp farming in coastal zone of Bangladesh and approaches for sustainable management. *International Journal of Environment Science Nature Research* 20 (3): 556038. https://doi.org/10.19080/IJESNR.2019.20.556038.

Sharma, N., and Nikolik, G. 2023. What to expect in aquaculture in 2024, Rabobank Research. https://www.rabobank.com/knowledge/q011418194-what-to-expect-in-aquaculture-in-2024.

Sherratt, R. 2020. Expert topic: Giant tiger prawn. https://www.aquafeed.co.uk/expert-topic-giant-tiger-prawn-22871/.

Stringer, C., B. Burmeister, and S. Michailova. 2022. Modern slavery and the governance of labor exploitation in the Thai fishing industry. *Journal of Cleaner Production* 371: 133645. https://doi.org/10.1016/j.jclepro.2022.133645.

Su, Y., H. Yua, C. Gaob, S. Suna, Y. Lianga, G. Liua, X. Zhanga, Y. Donga, X. Liua, G. Chen, H. Shaoa, A. McMinna, and M. Wang. 2024. Effects of vegetation cover and aquaculture pollution on viral assemblages in mangroves sediments. *Journal of Hazardous Materials* 476: 135147.

Sundall, E., J. Brijs, and A. Gräns. 2024. The quest for a humane protocol for stunning and killing Nile tilapia (Oreochromis niloticus). *Aquaculture* 593 (15): 741317. https://doi.org/10.1016/j.aquaculture.2024.741317.

Sylvester, E.V., B.F. Wringe, and S.J. Duffy. 2018. Migration effort and wild population size influence the prevalence of hybridization between escaped farmed and wild Atlantic salmon. *Aquaculture Environment Interaction* 10: 401–11. https://doi.org/10.3354/aei00277.

Thornber, K., D. Verner-Jeffreys, S. Hinchliffe, M.M. Rahman, D. Bass, and T.R. Tyler. 2019. Evaluating antimicrobial resistance in the global shrimp industry. *Reviews in Aquaculture* (2020) 12: 966–986.

Towers, L. 2012. Farmed Shrimp still safe, antibiotic use declining, The Fish Sit. https://thefishsite.com/articles/farmed-shrimp-still-safe-antibiotic-use-declining.

Uddin, M.H., M. Shahjahan, A.K.M.R. Amin, M.M. Haque, M.A. Islam, and M.E. Azim. 2016. Impacts of organophosphate pesticide, sumithion on water quality and benthic invertebrates in aquaculture ponds. *Aquaculture Reports* 3: 88–92.

UN. 2024. United Nations Sustainable Development Goal 6: Clean Water and Sanitation, United Nations. https://www.un.org/sustainabledevelopment/water-and-sanitation/.

Ungoed-Thomas, J. 2022. Farmed fish feel pain, stress and anxiety and must be killed humanely, global regulator accepts. The Observer, 8 October.

USDA. 2024. Production—Soybeans, US Department of Agriculture, Foreign Agricultural Service. https://fas.usda.gov/data/production/commodity/2222000.

van de Vis, H., W. Abbink, B. Lambooij, and M. Bracke. 2014. Stunning and killing of farmed fish: How to put it into practice. In *Encyclopedia of meat sciences*, 2nd Ed., 421–426. https://doi.org/10.1016/B978-0-12-384731-7.00199-9.

Wang, Y., Y. Zhao, X. Han, J. Wang, C. Wu, Y. Zhuang, J. Liu, and W. Li. 2023. A review of organophosphate esters in aquatic environments: Levels, distribution, and human exposure. *Water* 15 (9): 1790.

Williams, N. 2023. Occupational health in aquaculture, occupational medicine 73 (5): 293. https://doi.org/10.1093/occmed/kqad034.

Wood, G. 2015. From our prison to your dinner table, Pacific Standard, 3 March. https://psmag.com/economics/from-our-prison-to-your-dinner-table/.

Woodruff, M.L. 2018. The fish in the creek is sentient, even if I can't speak with it. Transformação 41 (spe): 119–152. https://doi.org/10.1590/0101-3173.2018.v41esp.08.p119.

World Bank. 2023. Global seaweed new and emerging markets report 2023. World Bank. https://openknowledge.worldbank.org/handle/10986/40187.

Wringe, B.F., N.W. Jeffery, R.R. Stanley, et al. 2018. Extensive hybridization following a large escape of domesticated Atlantic salmon in the northwest Atlantic. *Communication Biology* 1: 108. https://doi.org/10.1038/s42003-018-0112-9.

Wuertz, S., D. Bierbach, and M. Bögner. 2023. Welfare of decapod crustaceans with special emphasis on stress physiology. *Aquaculture Research.* https://doi.org/10.1155/2023/1307684.

WWF. 2024. Farmed Shrimp, World Wildlife Fund. https://www.worldwildlife.org/industries/farmed-shrimp.

Yuan, Y., Y. Yuan, Y. Dai, Z. Zhang, Y. Gong, and Y. Yuan. 2020. Technical efficiency of different farm sizes for tilapia farming in China. *Aquaculture Research* 51: 307–315.

Ziarati, M., M.J. Zorriehzahra, F. Hassantabar, Z. Mehrabi, M. Dhawan, K. Sharun, T. Talha Bin Emran, K. Dhama, W. Chaicumpa, and S. Shamsi. 2022. Zoonotic diseases of fish and their prevention and control. Veterinary Quarterly 42 (1):95–118. https://doi.org/10.1080/01652176.2022.2080298.

Zorriehzahra, M.J., and R. Banaederakhshan. 2015. Early Mortality Syndrome (EMS) as new emerging threat in Shrimp industry. *Advances in Animal Veterinary Science* 3 (2S): 64–72. https://doi.org/10.14737/journal.aavs/2015/3.2s.64.72.

CHAPTER 5

Palm Oil and Soya

Abstract This chapter analyses the environmental, social, and ethical challenges associated with the farming of key aquaculture species: tiger prawns, king prawns, Atlantic salmon, and tilapia. These species dominate global aquaculture, providing a crucial source of protein and economic livelihoods but also contributing to significant environmental degradation and social injustices. The chapter outlines the environmental impacts of aquaculture, including water pollution, habitat destruction, and the overuse of antibiotics, which exacerbate biodiversity loss and contribute to the emergence of antibiotic-resistant pathogens. Social issues such as labour exploitation, including instances of human trafficking and unsafe working conditions, are examined in the context of prawn and fish farming industries in regions such as Southeast Asia, South America, and Africa. The ethical implications of farming sentient creatures in intensive, high-density systems are also explored, with attention to welfare concerns and inadequate regulatory frameworks. The chapter concludes by highlighting the need for sustainable aquaculture practices and innovative alternatives to balance global food security, economic benefits, and ecological resilience.

Keywords Aquaculture · Antibiotic resistance · Biodiversity loss · Labour exploitation · Sustainable food systems

© The Author(s), under exclusive license to Springer Nature Switzerland AG 2025

S. Sneddon, *Food, Environmental Degradation and Injustices*, Palgrave Studies in Green Criminology,
https://doi.org/10.1007/978-3-031-87900-5_5

Introduction

The final substantive chapter moves away from species which are consumed more or less as they are, to consider Palm Oil and soya, which are ingredients of countless other processed foods as well as, in the case of palm oil, cosmetics, cooking fuels, and other commodities. Unlike the species examined in earlier chapters, these crops are central to vast supply chains, integrating industrial-scale agriculture, global trade, and multinational corporations. Palm oil and soya epitomise the relationship between economic development, environmental degradation, and ethical concerns, allowing an investigation of the challenges of balancing human progress with ecological and social justice.

The production of palm oil, predominantly in Indonesia and Malaysia, has become an emblem of industrial agricultural efficiency, capable of yielding high volumes of oil per hectare compared to other crops. Yet, this efficiency comes at a steep cost: the widespread deforestation of biodiverse rainforests, displacement of Indigenous communities, and significant greenhouse gas emissions. Similarly, the cultivation of soya, concentrated in Brazil, the United States, and Argentina, has driven deforestation in critical biomes such as the Amazon and Cerrado. Both industries exemplify the global inequalities inherent in modern agriculture, as economic benefits disproportionately favour corporations and distant consumers, while local and Indigenous communities bear the brunt of environmental and social harm.

The ethical implications of palm oil and soya production revolve around labour exploitation and the displacement of vulnerable populations. There is also an issue around corruption, and Sneddon (2021) demonstrated "the opportunities for criminal groups, corrupt public officials and corporate entities to engage in corrupt patterns of behaviour in relation to palm oil plantations". For palm oil, issues such as precarious contracts for workers and unsafe working conditions highlight systemic problems within the industry. In Malaysia, the dominance of foreign migrant labour on plantations raises additional concerns about community marginalisation and inadequate enforcement of labour rights. While the soya industry is more mechanised, reducing direct labour exploitation, its environmental impacts indirectly perpetuate social injustices, especially through land grabbing and deforestation that displace Indigenous and rural communities.

Environmental justice is central to this chapter, emphasising the unequal distribution of environmental risks and benefits. The clearance of tropical forests for palm oil and soya plantations significantly contributes to biodiversity loss, carbon emissions, and climate change. In Indonesia and Malaysia, palm oil plantations have replaced vast swaths of rainforest, leading to the loss of essential habitats and increased vulnerability for local communities. Similarly, the expansion of soya plantations in Brazil and Argentina has resulted in widespread deforestation, water pollution, and habitat destruction, with Indigenous populations disproportionately affected. Voluntary sustainability initiatives, such as the Roundtable on Sustainable Palm Oil (RSPO) and the Amazon Soy Moratorium (ASM), have achieved mixed results, highlighting both the potential and limitations of private governance in addressing environmental harm.

The ecological impact of palm oil and soya is multifaceted, encompassing habitat destruction, species endangerment, and pollution from agricultural run-off. Palm oil plantations in Southeast Asia have encroached on biodiverse ecosystems, threatening species like the orangutan, Sumatran rhinoceros, and Sunda tiger. Similarly, soya production in the Amazon has fragmented critical habitats, undermining ecosystem services and accelerating biodiversity loss. This section also considers the broader concept of species justice, examining the moral and ecological obligations humans hold to non-human life. The discussion is framed within the lens of intergenerational equity, emphasising the need to preserve ecological integrity for both current and future generations of humans and other species.

While palm oil and soya industries generate significant economic benefits, these are unevenly distributed. For smallholder farmers in Indonesia and Malaysia, oil palm cultivation can be a path out of poverty, providing steady income and enabling investments in education and healthcare. However, corporate entities dominate the supply chain, capturing the majority of profits. Similarly, in Brazil, large-scale soya farms benefit from economies of scale, while smallholders and Indigenous farmers face increasing marginalisation. The chapter considers the tension between economic growth and sustainability, and whether a shift to alternative crops or lab-based solutions could mitigate the industries' negative impacts without exacerbating poverty.

Overall, the chapter situates palm oil and soya production within a broader conversation about sustainability and global equity. It emphasises the need for systemic change, advocating for policies and practices that prioritise the well-being of marginalised communities and ecosystems while challenging the exploitative dynamics of global supply chains.

Environmental Justice

Palm Oil

The palm oil industry is predominantly centred in Indonesia and Malaysia but is starting to expand into South America (Colombia produced 1.9 million tons in 2023/4) and return to Africa (Nigeria produced 1.5 million tons in 2023/4) (FAS 2024a, b). Global production has grown steadily over the last decade from just over 60 million tons in 2013/4 to over 77 million tons in 2023/4, with the COVID-19 pandemic appearing to make little or no difference to overall production levels (FAS 2024a, b). Despite the continued but relatively modest rises over the past decade, the FAS predicted in 2020 that production could rise as high as 447 million tons by 2050 (FAS 2023). The Roundtable on Sustainable Palm Oil (RSPO) is a little more modest, but nonetheless predicted a rise to 150 million tons by 2050 (RSPO 2015). The growth in production requires commensurate growth in land use, which will put pressure on other land uses, from farming to housing to natural ecosystems. In Indonesia and Malaysia, for example Cramb and McCarthy (2016) say that the area cultivated for palm oil increased from 26,000 km^2 in 1990 to 150,000 km^2 in 2014. They also make the point that the oil palm industry is not one which is inclusive, and the allocation of resources and distribution of benefits is skewed "in favor of a powerful cross-country coalition of political, bureaucratic, and agribusiness interests" (Cramb and McCarthy 2016, p. 443). in the first decade of the twenty-first century, Lee et al. (2014) calculated that on the island of Sumatra, over 4,700 hectares of mangrove, over 380,000 hectares of peat swamp forest, and almost 290,000 hectares of lowland forest were lots to oil palm development (2014, p. 27) and Pye (2019) showed that almost "90% of this deforestation was caused by private palm oil corporations, including RSPO members, rather than smallholders" (2019, p. 219). The benefits of this development therefore head to the corporate oil palm companies, whereas the impacts of this destruction are keenly felt by the local communities which depend upon these forested areas for their livelihoods, cultural practices, food, and clean water.

The clearance of tropical forests for palm oil plantations significantly contributes to greenhouse gas emissions, driving climate change that disproportionately affects vulnerable populations with limited adaptive capacity (IPCC 2019). The IPCC report also states that enhancing this adaptive capacity in Indigenous and Local Knowledge (ILK) required formal education to be put in place as "these knowledge systems may become less relevant in certain areas where the rate of environmental change is rapid and the transmission of ILK between generations is becoming weaker" (2019, p. 755). McCarthy and Cramb (2009) have previously argued that far from seeing ILK as a worthwhile and valid resource, policies in Indonesia and Malaysia "tend to see the frontier in terms of absences: the lack of productive land uses, forms of modernised agriculture, and a disciplined labour force" (p. 113) and also that narratives "contrasting the problematic of 'backward' and 'unproductive' forms of agriculture, practised by undisciplined or 'lazy' local agriculturalists, with the desired form of rural modernity" (p. 113) have legitimised that approach.

Sheil et al. (2009) write that "Oil palm plantations apply large quantities of nitrogen-based fertilisers to plantations in order to increase and maintain yields" (p. 35). They also note that with the banning of the pesticide paraquat, which is an effective pesticide but also toxic to humans and non-human animals (Dinis-Oliveira et al. 2008), Malaysian palm oil producers switched to using glyphosate which, despite concerns over its toxicity to humans, has been approved by the European Food Safety Authority (Alvarez et al. 2023), and the US Environmental Protection Agency (EPA 1993). Interestingly, Mekhilefa et al. (2011) argue that palm oil is a promising biofuel crop (see below) purely because of the "low fertilizer, water and pesticide needed for the plantation" (2011, p. 1939). As we have seen in previous chapters, the risks associated with over use of pesticides and fertilisers are significant—from eutrophication and algal blooms in waterways to increased resistance in pest species. All of these will impact plantation workers and local communities more significantly than any other sector of the population. It is not just direct impact on the palm oil smallholders that causes problems. Rist et al. say that in Malaysia "Increasing rat populations associated with plantations have also decimated adjacent rice farms in some areas of West Kalimantan" (Rist et al. 2010, p. 1021).

Palm oil has impact not only on the health of those who work in the plantations, but on consumers too. The use of palm oil in foodstuffs is near ubiquitous, and the WWF-UK states that in the UK it is "in close to 50% of the packaged products we find in supermarkets, everything from pizza, doughnuts and chocolate, to deodorant, shampoo, toothpaste and lipstick" (2024). Sneddon (2021) suggests that the recent growth of palm oil usage in foodstuffs was partly triggered by drives to eliminate trans fatty acids from the human diet (2021, p. 10). A small number of sources link the consumption of palm oil to risks or heart and other cardiovascular disease and raised cholesterol (see for example Chen et al. 2011; Sun et al. 2015). Many more studies, however, suggest that palm oil consumption actually lowers these risks (see, for example, Loganathan et al. 2017; Adamu et al. 2024).

Palm oil as a biofuel is also facing some restrictions. In the European Union, the Renewable Energy Directive[1] requires that Member States phase out the use of palm oil (among other substances, including soy oil, see below) among other substances in biofuels by 2030, as it classifies palm oil among "high indirect land- use change-risk biofuels, bioliquids or biomass fuels produced from food and feed crops for which a significant expansion of the production area into land with high-carbon stock is observed" (Article 26(2)). In other words, it is not the carbon produced by the direct combustion of palm oil, which is the problem, rather the effects of forestry clearance in its production.

Overall, it is the initial forest clearance and production phases of palm oil which produce the largest environmental injustices, as those who reap the largest benefit are the large oil companies, and those who carry the heaviest burden are the already-marginalised Indigenous populations. Later in the chapter, we will discuss whether ceasing the commercial production of palm oil would bring to an end these injustices.

Soya

The US Department of Agriculture's Foreign Agricultural Service (FAS) reported that in 2023/24 three countries exported almost eighty percent of global soybeans—Brazil, at 153 million tons, the United States at 113 million toms and Argentina, at 48 million tons (FAS 2024b).

[1] Directive (EU) 2018/2001 OF THE EUROPEAN PARLIAMENT AND OF THE COUNCIL of 11 December 2018 on the promotion of the use of energy from renewable sources.

The expansion of industrial-scale soybean monocultures is closely linked to deforestation in the Amazon, Cerrado, and Gran Chaco regions, resulting in loss of ecosystem services, biodiversity, and traditional livelihoods (Henders et al. 2015; Fehlenberg et al. 2017). In this context, the unequal spatial and social distribution of environmental risks and benefits is most clearly exemplified in countries like Brazil and Argentina, where Indigenous communities, smallholder farmers, and rural peasants have faced disproportionate environmental and health burdens (Branford and Torres 2018). Branford and Torres also point out that on the positive side, Indigenous areas of the Amazon rainforest have lost less of their land than even the conservation, and that it is no coincidence that the "territories occupied by traditional populations are also known to be rich in biodiversity, with healthy and notably well managed forests" (p. 164).

Oliveira and Hecht (2016) however, point to the remarkable volte-face (which they call the "Amazon swerve" whereby the interests of agribusinesses have been articulated "in the discursive transformation of soy from a key driver of deforestation in the Amazon into the modern embodiment of the solution to the eternal problem in tropical development: combining environmental stewardship with development" (p. 268).

As forested areas in the Amazon are cleared to make way for soya plantations (and cattle farming, see Chap. 2), local and Indigenous communities lose both material and cultural resources. Their access to clean water, game, medicinal plants, and arable land is diminished, while agribusinesses and distant consumers benefit from cheap feedstock for livestock and other soya-derived products (Fearnside 2017). Fearnside also makes the point that:

> Even though the environmental services, such as avoiding global warming, recycling water, and maintaining biodiversity, are worth much more to human society than the money gained from destroying the forest, the institutional mechanisms needed to transform these services into a monetary flow and to use this flow to support the rural population without provoking perverse social effects are lacking (2017, p. 27)

The negative impacts of soy plantations have been abated to some extent by the existence of the Amazon Soy Moratorium (ASM). This was created in 2006 and involved soy farmers and environmental NGOs and "was the first voluntary agreement between civil society, industry and government with the aim of stopping the main soy traders from purchasing soybeans produced in areas of the Amazon cleared after July 2006" (Carvalho et al. 2019, p. 123). Although it has no legal basis, the ASM has been reasonably successful in stopping deforestation for soy plantations. Ziegert and Sotirov call the ASM a good example of "a hybrid governance mechanism that has been successfully running for nearly 20 years" (2024, p. 2). Fearnside (2017) and Carvalho et al. (2019) do sound a note of caution, however, pointing out that cattle ranchers may sell their land to soy farmers, thus meeting the requirements of the ASM since the land was already cleared. The ranchers may then purchase cheaper land elsewhere to clear for pasture, since there is no equivalent moratorium in place for cattle.

In addition to the impacts on Indigenous and marginalised communities caused by deforestation and land grabbing in both palm oil and soya industries, there are issues around unfair labour practices. In a lengthy investigation into labour practices in the Indonesian and Malaysian Palm Oil industry, Amnesty International (2016) found "harvesters [working] long hours, in excess of the limit of 40 hours per week set out under Indonesian law" (p. 8), a "lack of protective equipment for those involved in spraying pesticides" (p. 9), and criticised the RSPO for "not doing more to prevent this in its certified member producers" (p. 11). Pye (2019) too identifies problems in the Palm Oil industry, saying that in "Malaysia, work in the plantations and mills is performed almost wholly by migrant workers, whose basic rights are curtailed" (2019, p. 222). McBride (2024) says "95 percent of the palm oil labor force in Malaysia is foreign" (2024, p. 2). This practice of employing migrant workers has also proven problematic, and many "communities resent the fact that companies, despite promising them jobs upon which basis they agreed to the oil palm development, do not prioritise them in terms of training and/or employment in the plantations, preferring instead to bring in workers from outside, who, the companies state, are more qualified and experienced" (Colchester and Chai 2013, p. 385).

The soy industry is far more mechanised than the palm oil industry. For palm oil, almost all harvesting is done by hand, whereas the soy industry "both harvest and planting are carried out without human interference" (ONG Repórter Brasil 2008, p. 16). The reduction in "human interference" means that the opportunities for poor working conditions and other forms of labour exploitation are limited. Indeed, in their systematic review of the soy industry, Dreoni et al. (2022) did not find any significant evidence of unfair labour practices. The negative impacts they recorded were focused partly on the impact of deforestation and the associated pollution on drinking water, but mainly "more intangible dimensions such as freedom of choice and cultural value are found to be negatively affected" (2022, p. 1).

Voluntary certification schemes like the Roundtable on Sustainable Palm Oil (RSPO) have attempted to curb the worst abuses, but critics argue these measures are insufficient and have not meaningfully halted deforestation or safeguarded critical habitats (Pichler 2013). Pichler argues that the RSPO is flawed to its core, as "96% of the 638 RSPO members are private companies whereas only 4% are environmental or development NGOs" (2013, p. 377). Workers in palm oil plantations do not even have the possibility to join the RSPO, "which means that the interests of plantation workers are—if at all—represented by either the palm oil companies or an environmental or development NGO but not by themselves" (2013, p. 381). In 2024, the RSPO "Ordinary Membership" category (the only type of membership with voting rights and the permission to publicly acknowledge membership) is only open to organisation which "Purchase, use or trade ≥ 500 metric tonnes of palm oil" (RSPO 2024), so it appears as though Pichler's criticisms from over a decade ago have not borne fruit.

Overall, the production of palm oil and soybeans does seem to demonstrate a pattern of environmental injustice. While global demand fuels economic growth and corporate profits, it simultaneously undermines the well-being of marginalised communities, endangers biodiversity, and exacerbates climate change. This is far more prevalent in the palm oil industry than in the soy industry, where the ASM has provided partial relief. Indigenous knowledge, crucial to sustainable resource management, remains undervalued, and local communities bear disproportionate burdens. If both industries were to cease, there would be direct benefits to Indigenous populations, and those exploited in the production processes as well as wider indirect benefits from reducing deforestation. The next section will focus on the ecological and species impacts of both industries.

Ecological and Species Justice

As we saw above, palm oil cultivation often involves clearing biodiverse rainforests and draining carbon-rich peatlands (Lee et al. 2014) and this fundamentally alters the regional ecology. Yale University School of the Environment's Global Forest Atlas says that between "1990 to 2005, over 50% of new oil palm plantations in Malaysia and Indonesia were established through conversion of lowland forests" (GFA 2024) and this is a significant part of the environmental cost. It is not just in these areas where deforestation occurs for the palm oil industry. Vijay et al. (2016) looked at palm oil plantations in 20 countries across the world, and their results "show high rates of forest loss for palm oil production across a range of countries and continents, raising concerns about future expansions of oil palm plantations" (2016, p. 17).

The shift from old-growth tropical forests to monoculture plantations diminishes ecological integrity, meaning that forests which once supported intricate networks of plants, animals, and microbes are replaced with homogenised landscapes. These plantations lack the complexity and resilience of natural ecosystems, hosting fewer species and providing fewer ecosystem services (Foster et al. 2011). Areas which used to house both iconic and less well-known species (such as Orangutan, Sunda Tiger, Sumatran Rhinoceros, Sumatran Elephant, Borneo Pygmy Elephant, Sunda Pangolin, Malaysian Sun Bear, Sunda Clouded Leopard, Irrawaddy Dolphin, Javan Blue-Banded Kingfisher, and Nordmann's Greenshank) are being cleared, with a dramatic impact on biodiversity and the continued existence of some CITES Red-Listed Species. Although they do not overtly link the killing of Orangutan with forestry clearance for palm oil plantations, Sherman et al. (2022) make the point that "Orangutan-related crime in Central Kalimantan declined notably between 2007 and 2008, following extensive clearing for oil palm and associated human orangutan conflicts prior to 2007" (2022, p. 4). On the one hand, this may look like positive news, as any reduction in human-wildlife conflict is to be welcomed. However, the rationale actually appears to be that once the forest is cleared, the surviving Orangutan leave the area, which is much less positive.

One of the most glaring ecological impacts of soya production lies in the clearance of biodiverse habitats, notably the Amazon rainforest, the Brazilian Cerrado, and the Gran Chaco region spanning Argentina, Paraguay, and Bolivia. Industrial-scale monocultures often replace complex, species-rich habitats with large uniform fields dominated by a single crop. This land-use change fragments landscapes undermines ecosystem functions, and compromises ecological integrity (see, for example, Henders et al. 2015; Fearnside 2017). Furthermore, deforestation associated with expanding oil palm plantations—often to accommodate increasing waste treatment or storage areas—amplifies biodiversity loss. Habitat fragmentation, which may leave areas of "High Conservation Value Forest" (HCVF) surrounded by areas of monoculture, diminishes the resilience of tropical forests, and marginalises species unable to adapt to altered landscapes (Pye 2019, p. 219).

Maluin et al. (2020) have identified a potential problem with palm oil production. The irony is that climate change (of which deforestation for palm oil plantation is a major driver) may be causing a reduction in both fruit quantity per tree, and fruit quality, leading to lower oil extraction rates (2020, p. 2). This means that in order to generate the same quantity of oil, more trees will be needed, and thus both more space and more waste.

Less than one fifth of palm oil has some sort of sustainability certification (mostly from the RSPO), and this often overlooks the origins of the plantations—in effect labelling as sustainable a process which committed all of its unsustainable activities in the past. Even with this labelling, Pye (2019) points out that "Buying the commodity marketed as "sustainable palm oil" might encourage a corporation to improve the management of their monoculture plantations, but it does not address the problem of the expanding area of monoculture plantations" (p. 220).

The ecological impact of waste generation in the palm oil industry is both far-reaching and multifaceted, affecting soil quality, water resources, greenhouse gas (GHG) emissions, and biodiversity. Maluin et al. (2020) note that the oil produced in a palm oil plantation only amounts to only around 10% of the biomass, whereas the waste generated accounts for 90% (2020, p. 3). They do note, however that "this waste can be utilized as sustainable and renewable biomass resources and can be used in biocomposites, paper production, construction board filler, power generation, soil wood, etc." (2020, p. 5). Although this is possible, it is not yet commonplace, however. Once the palm fruits are harvested, they need to be processed and, even though the techniques have improved over the years, one ton of palm oil results in roughly three tons of "POME" (Palm Oil Mill Effluent, the waste water which is often discharged untreated into the sea) (Hassan et al. 2005). In addition, the process produces waste "which ranges from oil palm empty fruit bunches (OPEFB), oil palm frond (OPF), oil palm shells (OPS), oil palm mesocarp fiber (OPMF), oil palm trunk (OPT), palm oil fuel ash (POFA)", (Maluin et al. 2020, p. 5). However, POME, is one of the most ecologically problematic wastes due to its high biochemical oxygen demand (BOD) and chemical oxygen demand (COD) (Zaied et al. 2019; Lokman et al. 2021).

Pristine rivers will have a BOD of under 1 mg/l (Li and Liu 2019), moderately polluted rivers between 2 and 8 mg/l, and severely polluted rivers over 8 mg/l (2019, p. 214). The BOD of untreated POME is 25,000 mg/litre (Ahmad et al. 2003), which puts it 3,000 times higher than the standard for a severely polluted river. In terms of the chemical oxygen demand, George et al. (2010) suggest a safe value of 10 mg/l, but the Environment Agency in England has two sets of limits, those on what it calls a "Look Up Table" (LUT)[2] and Maximum Compliance limits. The LUT limit for COD is 125 mg/l and the maximum compliance limit is 250 mg/l, and the COD for untreated POME is 50,000 mg/litre (Ahmad et al. 2003). In addition, Zaied et al. (2020) show that "POME contains several suspended and dissolved elements that can produce unpleasant odors after organic breakdown" (2020, p. 2).

Wu et al. (2010) show that the waste from palm oil production produced significant quantities of methane (CH_4) which is an even more problematic Greenhouse Gas than carbon dioxide. They suggest that anaerobic digestion treatment for POME would allow this methane to be harvested and sold alongside biogas, which is already harvested (p. 1481). Yacob et al. (2006) agree and have calculated that 12.36 kg of methane can be produced by the anaerobic digestion of one ton of POME (p. 194). However, "CH4 is not harvested in most of the palm oil mills due to the lacking of infrastructure to channel this excess energy" (Wu et al. 2010, p. 1486). Yusoff (2006) too explores the use of waste in the palm oil industry in Malaysia, saying that "EFB and POME have been used extensively as mulch and organic fertilizer in oil palm areas while palm fibre and shell are used as fuel, making the palm oil mill self-sufficient in energy" (p. 88).

In some cases, the ecological impacts of palm oil industry waste extend beyond the immediate vicinity of plantations and mills. Nutrient and sediment run-off from POME sites may affect downstream ecosystems, polluting estuaries and coastal areas, and harming species reliant on clean water and stable habitats (Okereke and Ginikanwa 2020).

Yu et al. (2002) show that processing soybeans into bean curd produces effluent waste water with a COD of "10-20 g/l" (p. 507), which is 10,000–20,000 mg/litre. While this is highly polluted, Su Hongyang et al. (2011) demonstrated that this wastewater could be

[2] The LUT permits the limit to be exceeded a number of times in a rolling 12-month period, depending on how often samples are taken in that period (EA 2019).

successfully cleaned up by almost 80% by using a strain of algae called Chlorella pyrenoidosa. Qiu et al. (2019) used two different types of Chlorella algae and suggested that in addition to improving the quality of the wastewater, the algae could prove to be useful biofuel.

Soybeans only produce around half of the nitrogen they need (Mosaic 2024) and will need additional nitrogen-based fertiliser, as well as Potassium, Phosphorous, Sulphur, Zinc, Boron, and Copper (Mosaic 2024) to ensure good cropping levels. As with any crop which require additional fertiliser, there is a significant risk of run-off, and of the fertiliser reaching surface water. Elevated nutrient loads in surface waters can promote algal blooms, deplete dissolved oxygen, and reduce aquatic biodiversity. Moreover, leaching of agrochemical residues into soil and groundwater can disrupt soil microbial communities, reduce soil fertility, and pose health risks to wildlife and humans (Qi et al. 2023).

It is not just water quality which can be negatively impacted by waste products from the soybean processing industry. Air quality can also be impacted through the burning of residual biomass (Stone et al. 2010). Agricultural fires set to clear land or dispose of waste produce carbon dioxide (CO_2), nitrogen oxides (NOx), and particulate matter (PM), contributing to climate change and local air pollution (Reddington et al. 2021). The Amazon and Cerrado regions, for instance, have experienced substantial deforestation and associated biomass burning partly linked to soy expansion, and the associated emissions degrade regional air quality and ecological integrity (Butt et al. 2020).

The ecological injustices of soya production are deeply intertwined with the global climate crisis. Large-scale deforestation releases vast amounts of carbon stored in forests, peatlands, and savannas, accelerating climate change (Henders et al. 2015; IPCC 2019). Additionally, the loss of vegetation cover reduces rainfall recycling and amplifies regional droughts, further stressing ecosystems and species already stretched to their adaptive limits (Nepstad et al. 2014). In 2024, the Brazil suffered the highest level of forest fires for twenty years, impacting the Amazon rainforest, the Cerrado region, and the globally important Pantanal wetlands in the State of Mato Grosso do Sul. In total, over 400,000 km^2 of forest was lost (GWIS 2024).

In terms of associating palm oil and soybean production with ideas of species justice, both crops are so widespread in their geographical reach that focusing on one or two specific species which are negatively impacted would be unrepresentative, and since deforestation plays such an integral role in both crops, it would repeat many of the points raised in Chap. 2. Instead, we can consider whether the idea of intergenerational equity can be morphed successfully with species justice. The concept of intergenerational equity in environmental ethics and international law was pioneered by Brown Weiss (1989) and is part of the movement from which emerged broader ideas of sustainable development in the 1990s. Weiss' concept, which owed a debt to a lot of Indigenous cultural beliefs, was that the present generation hold the Earth and the Earth's resources on trust for future generations. In a later expansion of the concept, Brown Weiss set out three normative principles of intergenerational equity (1992, pp. 22–23):

- First, conservation of options—conserving the diversity of the natural and cultural resource base so as not to limit the options of future generations;
- Second, conservation of quality—each generation must maintain quality, so the planet is passed on in as good as, if not better a state that it was inherited; and
- Third, conservation of access—to the legacy of past generations.

This concept goes further than the Brundtland Commission's 1992 definition of Sustainable Development which it influenced and speaks to cultural and societal values as well as environment quality. What both share is their focus on generations of humans—neither was designed to be applied directly it to other species of animal, let alone more widely. However, it is easy to see a conceptual link to species justice, and to reframe the idea and to cover human society maintaining conservation of options, quality, and access in relation to non-human species. By applying Weiss's intergenerational lens to non-human species, we acknowledge that current generations hold Earth's biodiversity in trust, not only for future humans but also for the future flourishing of non-human life.

We have seen that both palm oil and soybean production have led to widespread environmental and ecological harm, and that neither would satisfy the requirements for either species justice or intergenerational equity. By incorporating species justice into the intergenerational equity framework, we can begin to acknowledge that obligations to the future cannot be merely anthropocentric; they must ensure that future beings—human and non-human—inherit a rich tapestry of life and ecological resilience.

Overall, we can see that the expansion of palm oil and soybean plantations has produced notable ecological and species injustices, aligning with White's (2013) analysis of environmental harms that marginalise non-human life and ecosystems. Just as Edith Brown Weiss's intergenerational equity highlights our obligations to future human generations, White's frameworks underscore the moral and legal responsibilities we owe to non-human species and ecological communities. Halting commercial production might initially seem beneficial, as it would remove pressures that drive habitat loss and endangerment of species, as well as improving water and air quality. However, an abrupt cessation could bring complex socio-economic repercussions, potentially harming communities dependent on these industries and creating new injustices. These will be explored in more depth in the following section.

Economic

Palm Oil

The inherent conflict embedded within the palm oil industry is neatly illustrated by Sheil et al. (2009), who say that:

> Oil palm expansion can contribute to deforestation, peat degradation, biodiversity loss, forest fires and a range of social issues. But oil palm is also a major driver of economic growth and a source of alternative fuel (p. 1)

As we have seen, despite the species not being a native to the region, Malaysia and Indonesia have proven to be ideal growing conditions for the oil palm, and it is unsurprising that the two nations produce around 90% of the world's palm oil (IndexMundi 2024). It is an important income generator for both countries, but as with the other products discussed in previous chapters, the foreign currency benefits of exporting palm oil (in whatever format) are only part of the economic story. Voora et al. (2020) have estimated that the industry employs 3 million smallholder farmers and 2.9 million downstream jobs so the impact on the wider economy must not be underestimated.

However, one of the reasons oil palms have been so avidly embraced as crops is that they are very efficient in terms of the oil generated per hectare, producing 3.3 tons of oil, compared to 700 kg for each of Coconut, Sunflower, and Rapeseed (Canola) (Statista 2020). This means that if the almost 200,000 km^2 of palm oil plantations across Indonesia and Malaysia were to be simply "swapped out" for sunflower plantations, an additional 750,000 km^2 of land would be required[3]—that is an area twice the size of Germany. Furthermore, oil palms will reach maturity and start to produce fruit after two to three years and will continue to do so for around thirty years (Poku 2002). Coconut palms on the other hand take around a decade to reach maturity, though they will then continue to produce fruit for 60–70 years (Gordon and Jackson 2017). Sunflowers and rapeseed are annual crops, so will need replanting and harvesting annually.

In 2023/4, Indonesia exported 22.2 million tons of palm oil, which was a ten percent drop on the previous year (Rahmanulloh 2024, p. 5). In Malaysia, which has suffered abnormally high temperatures and low rainfall in the early part of 2024, exports were 16.2 million tons, but this is expected to fall in 2024/25 (McBride 2024, p. 6) partly as a result of the import tax hike by India from 5.5% to 27.5% for crude palm oil and from 13.75% to 35.75% for refined palm oil (p. 4).

Smallholder farmers, who often work under contract farming arrangements or own their own plots, can experience significant income improvements when transitioning from subsistence agriculture to oil palm cultivation. This shift may allow them to invest in education for their children, healthcare, and housing, including buying plots of land to give their children on marriage (Li 2015, p. 6) thereby improving long-term socioeconomic well-being. This income would have to be replicated without palm oil, and Li (2015) also found that smallholders who were wholly dependent upon oil palm for income said they needed 6 hectares of palm oil plantation to survive—so using the calculations above, that would equate to 28 hectares if the oil palms were replaced by sunflowers, rapeseed, or coconut.

Rist et al. (2010) warn that volatility in global trade can cause prices to rise and fall, and that while larger corporate palm oil producers would be

[3] Based on "Area Harvested" figures in Rahmanulloh (2024) and McBride (2024).

able to weather such storms, the impact would be amplified for smallholders. However, that also note that in "Malaysia, the incidence of poverty among oil palm smallholders has been negligible since the early 1980s, compared to small-scale producers of other agricultural commodities such as fish and rice" (Rist et al. 2010, p. 1019). Ending palm oil production would thus seem very likely to have a much more significant negative economic impact on smallholders than on corporate producers.

Soya

There is an interesting parallel between the oil palm, which originated in Africa and for which the global leaders are now in Southeast Asia. The soybean (*Glycene max*), originated in East Asia, and the global leaders in terms of production are now in the Americas. Though this chapter has largely focused on soybeans as a human food crop, but as Fraanje and Garnett (2020) point out, only six percent of soy is processed directly for human food (as tofu, tempeh, soy milk, edamame beans, etc.), and a further 13% where the soy oil is a constituent part of human food.[4] While some soy (around three percent) is used as a biofuel, the remaining 80% is used for animal feed (Fraanje and Garnett 2020, p. 5). Linking back to the discussions in Chap. 2, if human consumption of animals, especially pigs, was to be halted or significantly reduced, much of the protein could potentially be replaced by the soy which is currently used as animal feed.

As with palm oil, soy is a major export commodity. In terms of financial value, Brazil (US$53bn) and the United States (US$28bn) are far and away the largest exporters, and China (US$59bn) imports 66% of soya beans (TrendEconomy 2024). In terms of the contribution soybean exports make to a country's total export, 20% of Paraguay's exports are soy, as are 15% of Brazil's. Since Brazil is the largest producer and exporter of soy, this section will focus on the economic impact it has in the country, and how this is distributed. As with palm oil, the macro-economic impacts of global trade are only the headline figures, the arguably more impact is at the granular, smallholder level.

The State of Mato Grosso, which borders Bolivia, is the largest producer of soybeans in Brazil, accounting for around 28% (IPAD 2024), with 54% of all agricultural land laid down to soy (Zhang and Chen

[4] 87% is used for processing. Of that, 19% is processed into oil, and of that 80% is for human consumption. Source: Fraanje and Garnett (2020, p. 5).

2021). Carvalho et al. (2024) found that while 80% of soy farms where deforestation had taken place were small (under 400 hectares) or medium (400–1500 hectares), the 20% of large farms had significantly more soy planted, suggesting significant economies of scale. Zhang and Chen (2021) found that the average soy farm size in Mato Grosso was 500 hectares, compared with around 70 hectares in other Brazilian States. As we know, the ASM has been in place for nearly two decades, so illegal deforestation is on the wane; however, Carvalho et al. also found that "most land illegally deforested was in large soy farms" (2024, p. 7). The industry in the State is also dominated by a few large companies including Amaggi Group, which is "one of the largest private producers of soybeans in the world" (WBA 2024) and which Zhang and Chen say "had a monthly net income of over $126.8 million USD in 2016 versus a typical soy farmer who earns approximately $210 USD or $1000 reals" (2021).[5] Interestingly, Statista (2023) lists the average income of employed people in Mato Grosso as around R$3,000 per month or triple the average income of soy farmers.

However, as with palm oil, there is conflicting evidence about the impact of soy farming on poverty. In Paraguay, Guereña (2013) argues that the move to soy monoculture has exacerbated poverty, so that "almost half of rural families live below the threshold of poverty, and their income does not cover 44.4% of the cost of the basic food basket" (2013, p. 7). However, rather than being the result of anything inherent within soy, this has happened as a result of unequal land distribution where "an elite of large land owners ('latifundistas') holds almost all of the land available for agriculture and livestock, while the vast majority of small-scale, family farmers lack sufficient land for production" (2013, p. 8). Piras et al. (2021), who focus on soy production in Brazil acknowledge that "soy expansion boosts local income because it is an important economic activity" (2021, p. 6) but add the caveat that it "may also intensify income inequality" (p. 6). Weinhold et al. (2013) are altogether more, and "find that increases in soy cultivation are associated with strong increases in rural median household income and rural and urban GDP per capita" (p. 142).

The palm oil and soya industries are both good examples of the tensions which exist between economic growth and environmental

[5] There is no indication that the Amaggi Group is involved in any way with illegal activities.

stewardship. Palm oil, largely concentrated in Indonesia and Malaysia, supports millions of smallholder farmers, and provides significant export income. Oil palms are highly efficient, yielding more oil per hectare than alternatives seed oil crops, meaning any shift away from palm oil could require substantially more land. On the other hand, palm oil cultivation involves deforestation, loss of biodiversity, and ecological harm. While smallholders have benefitted economically, ceasing production would likely have severe financial consequences for them, though corporate producers could better weather such changes. Soy too generates vital export revenue and may improve local incomes, but this is often at the cost of exacerbating inequality and environmental degradation.

Future

Climate change is expected to have profound impacts on all palm oil-producing regions, both in the humid tropics of Southeast Asia, as well as increasingly in Africa and Latin America. Shifts in rainfall patterns, more frequent droughts, and heightened flood risks are all likely (IPCC 2019). Murphy et al. (2021) identify that climate change is one of the significant future threats that face the oil palm industry. Maluin et al. (2020) have identified that climate change may be causing a reduction in both fruit quantity per tree, and fruit quality, leading to lower oil extraction rates (2020, p. 2). Abubakar et al. (2023) say that "Heat stress further exacerbates the situation, as it diminishes photosynthesis, increases respiration rates, and induces water stress and premature aging in oil palm trees" (p. 95). This means that in order to generate the same quantity of oil, more trees will be needed, and thus both more space and more waste. However, Fleiss et al. (2018) predict that as a result of increased heat and drought there will be 75% less land suitable for growing oil palms by 2100, and that this will not be compensated for by new potential growth areas (2018, p. 5).

Although not directly threatened by rising sea levels, the additional pressure on agricultural land caused by climate migrants will potentially lead plantations to exploit yet more of the forested areas. Parts of the Indonesian capital Jakarta, for example, which is home to over 30 million people are sinking by up to 20 cm a year, as a result of illegal water abstraction (Renaldi 2022). This sinking, combined with sea level rise of 3.3 mm a year, means that "as much as 40% of the city now sits below sea level" (NASA 2021).

Unlike many other foodstuffs, there is currently little sign of any plans to genetically modify palm oil. Indeed, the Malaysian Palm Oil Council (the trade body for Malaysian Palm Oil) makes a point of saying that "Palm oil is a 'balanced' oil that is free of cholesterol, trans fats, and genetically modified organisms (GMO)" (MPOC 2016). However, while the Malaysian Palm Oil Board (the government department responsible for promoting palm oil) agrees that there is no GM palm oil at present, it confirmed that "GM is part of the research that we have" (MM 2023) and that the focus of GM research is on improving yield.

There is also some research on "lab-grown" palm oil, which is generating interest. American company, C16 Biosciences, has produced "Palmless" a range of products which are produced in a laboratory using "naturally occurring yeast" (Palmless 2024), and the base oil was certified by the USFDA (US Food and Drink Administration) for use in cosmetics in 2023. A second US Company, Xylome, also produces through yeast fermentation "a palm oil substitute that is sustainable, ethical, and affordable" (Xylome 2024) which also received USFDA certification in 2023. Other companies are currently experimenting with lab produced palm oil substitutes, and it is likely that such products may help reduce reliance on oil palms in the future.

Soya

One of the reasons soy as a crop is in such demand is, as stated earlier, its versatility and ability to be used as human food, animal food, and as biofuel. This may also be its weakness—as demand for biofuels increases, the pressure on soy as animal feed will increase, and is likely to lead to greater deforestation, regardless of any moratoria. This will exacerbate the impact of climate change, which is already set to impact on global soy production, and as the climate changes and the range of pests and diseases adapt, soya producers too will have to find ways of mitigating the impact. Tomasz et al. (2024) note that in the area of Argentina which they studied "Rainfall is the main input since the production across the implanted area of 17 million hectares is carried out without irrigation. Therefore, there is total exposure to climate risk, especially to rainfall levels and temperature" (2024, p. 2).

Unlike palm oil, soybeans have been being genetically modified for nearly thirty years, with Monsanto releasing a glyphosate-resistant plant (GTS 40-3-2, often called "Roundup Ready Soy" in 1995. Since then, different genetic modifications have been made to soy, to improve resistance to pests, to improve yield and so on, and in 2020, "GMO soybeans made up 94% of all soybeans planted" in the United States (USFDA 2024). The USFDA also state that "Most GMO soy is used for food for animals, predominantly poultry and livestock, and making soybean oil" (2024). In 2023, a biotechnology company in the UK produced "Piggy Sooy" a transgenic GM soy plant where "one quarter of the soluble proteins are pig proteins" (Le Page 2023). The intention is to reduce some of the issues that are currently associated with lab-grown meat.

Much as soy is grown as a feedstuff for animals, Ben-Arye et al. (2020) have used textured soy protein as the scaffolding to grow "Cell-based meat" (also called "lab-grown meat" see Chap. 2), and notes that doing so achieved "the goal of replicating the sensation and texture of a meat bite" (p. 210). Unlike the lab-grown palm oil sector, there appears to be little drive for a lab-grow soy sector. Davis et al. (2024) suggest that unlike palm oil, where the main product is the oil "substantial demand for soy protein (for example, meal for animal feed) co-produced with soybean oil might limit replacement of soybean oil, unless demand for soy protein could be reduced in parallel to synthetic oil-driven decreases in demand for soybean oil" (2024, p. 93). They suggest that reduced consumer demand for chicken and pork will reduce this demand for soy protein and improve the potential for lab-based soy oil.

Overall, we can see that both the palm oil and soybean industries face pressing sustainability challenges as climate change intensifies. For palm oil, shifting rainfall patterns, rising temperatures, and drought threaten yields, potentially driving further expansion into forests to maintain output, worsening environmental and biodiversity impacts. While lab-grown substitutes and genetic modification could alleviate some pressures, they currently remain somewhat niche solutions. Soy, meanwhile, is embedded deeply in global supply chains, primarily as animal feed, and genetic modification has long been a key strategy to maintain productivity. Yet, reducing dependency on livestock products could curtail soy demand. Ultimately, integrated approaches—balancing innovation, responsible consumption, and equitable policies—are required.

Conclusion

The chapter has examined the various effects of the palm oil and soya industries, which highlight the conflict between economic growth, environmental protection, and social fairness. These crops symbolise global agricultural systems—they serve vast supply chains and consumer needs, but their cultivation also creates significant ethical, ecological, and economic dilemmas.

The palm oil and soya sectors present serious ethical issues, especially regarding their treatment of workers and the uprooting of marginalised communities. Labour exploitation, unstable contracts, and insufficient safety protocols continue to exist in the palm oil sector, particularly in Malaysia and Indonesia, where the workforce mainly consists of migrant labourers. In the soya sector, although mechanisation diminishes direct exploitation, practices like land grabbing and deforestation continue to indirectly foster social injustices, with Indigenous communities bearing a disproportionate impact. Corruption worsens these challenges, allowing corporations to focus on profits instead of fair treatment and care for the environment. Although voluntary certification initiatives such as the RSPO and the Amazon Soy Moratorium have made progress in tackling these issues, deep-rooted disparities continue to persist.

Environmental justice highlights the uneven distribution of environmental advantages and drawbacks. The clearing of diverse rainforests for palm oil and soybean farms has extensive repercussions, such as habitat destruction, increased greenhouse gas emissions, and climate change. The communities reliant on these forests for cultural and material resources bear the heaviest consequences, while corporations and distant consumers enjoy the rewards. Palm oil farming in Southeast Asia has resulted in significant deforestation, with the environmental repercussions far exceeding the short-term economic benefits. Likewise, soya farming in South America has ravaged the Amazon, Cerrado, and Gran Chaco biomes, worsening biodiversity decline and regional climate challenges. While the establishment of certification programmes has offered some relief, it falls short of addressing the magnitude of damage inflicted by these industries.

The ecological consequences of palm oil and soya farming are alarming. Both agricultural practices replace diverse ecosystems with single-crop fields, leading to a decrease in biodiversity and a weakening of natural habitats' resilience. The likely extinction of species such as orangutans and tigers in Southeast Asia and the habitat fragmentation in South America highlight the severe effects on wildlife. We have also seen how it is possible to connect ecological challenges to the broader notion of species justice, positing that humanity bears a moral duty to protect biodiversity for the sake of future generations. By situating species justice within the framework of intergenerational equity, the chapter underscores the necessity of safeguarding ecological integrity not only for human interests but also for the inherent worth of non-human organisms.

From an economic perspective, the palm oil and soya sectors offer substantial advantages, especially for smallholder farmers and national economies that depend on exports. Palm oil cultivation supports millions of jobs in Indonesia and Malaysia, providing a means of escaping poverty for numerous individuals. Similarly, soya serves as an essential export for countries such as Brazil and the United States, enhancing local livelihoods and contributing to national economic growth. Nonetheless, these advantages are not shared equally. Large corporations control the supply chain, reaping most of the profits while sidelining small farmers and Indigenous populations. Additionally, the economic fluctuations of global commerce and the environmental impacts of production intensify these disparities.

The question of whether an immediate end to commercial palm oil and soya production would bring net benefits or harm is complex. It could theoretically alleviate environmental destruction, reduce greenhouse gas emissions, and improve species conservation. It might also dismantle exploitative labour practices and curtail the displacement of Indigenous communities, although it is likely that these practices will continue, but in respect of replacement crops.

However, the end of the trade could also have significant adverse consequences. For millions of smallholder farmers, palm oil cultivation represents their primary source of income. Abruptly ending production without appropriate replacement sources of income could plunge these communities into poverty, exacerbate inequality, and lead to socio-political instability. The loss of palm oil would also require alternative crops, which may demand more land and resources, potentially exacerbating environmental degradation. Soya presents similar dilemmas. A ban would mitigate deforestation and habitat loss in critical biomes, but it would disrupt global food systems, particularly the livestock industry, which relies heavily on soya as feed. This could lead to increased food prices and economic hardship for producers and consumers alike.

Rather than a complete and immediate cessation therefore, a more nuanced approach is essential. Policies that incentivise sustainable practices, such as stricter enforcement of certification schemes and increased support for Indigenous land rights, could mitigate many of the harms associated with these industries. Technological innovations, including lab-grown substitutes and genetic modifications, may offer additional pathways for reducing environmental impact. Reducing global reliance on palm oil and soya through shifts in consumption patterns—such as decreasing demand for processed foods and livestock products—could also play a critical role. Public awareness campaigns and international cooperation will be crucial in driving these changes.

Sources

Abubakar, A., J. Gambo and M.Y. Ishak. 2023. Chapter Three—Navigating climate challenges: Unraveling the effects of climate change on oil palm cultivation and adaptation strategies in Cohen MJ (ed) Advances in Food Security and Sustainability, Elsevier. https://doi.org/10.1016/bs.af2s.2023.07.002.

Adamu, M.A., E.S. Almeida, C.R. Carneiro, C.A.S. da Silva, S. Monteiro, J.S.D.R. Coimbra. 2024. Palm Oil (Elaeis guineensis): A journey through sustainability, processing, and utilization. *Foods* 13 (17): 2814. https://doi.org/10.3390/foods13172814.

Ahmad, A.L., S. Ismail, and S. Bhatia. 2003. Water recycling from palm oil mill effluent (POME) using membrane technology. *Desalination* 157 (1–3): 87–95. https://doi.org/10.1016/S0011-9164(03)00387-4.

Álvarez, F. et al. 2023. Peer review of the pesticide risk assessment of the active substance glyphosate. *EFSA Journal* 21 (7): e08164. https://doi.org/10.2903/j.efsa.2023.8164.

Amnesty International. 2016. The great palm oil scandal: Labour abuses behind big brand names. https://www.amnesty.org/en/wp-content/uploads/2021/05/ASA2151842016ENGLISH.pdf.

Ben-Arye, T., Y. Shandalov, S. Ben-Shaul, et al. 2020. Textured soy protein scaffolds enable the generation of three dimensional bovine skeletal muscle tissue for cell-based meat. *Nat Food* 1: 210–220. https://doi.org/10.1038/s43016-020-0046-5.

Branford, S., and M. Torres. 2018. *Amazon besieged: By dams, soya, agribusiness and land-grabbing*. Rugby: Practical Action Publishing.

Brown Weiss, E. 1989. *In fairness to future generations: International law, common patrimony, and intergenerational equity*. Transnational Publishers, United Nations University.

Brown Weiss, E. 1992. In fairness to future generations and sustainable development. *American University International Law Review* 8 (1): 19–26.

Butt, E.W., L. Conibear, C.L. Reddington, E. Darbyshire, W.T. Morgan, H. Coe, P. Artaxo, J. Brito, C. Knote, and D.V. Spracklen. 2020. Large air quality and human health impacts due to Amazon forest and vegetation fires. *Environmental Research Communications* 2 (9). https://doi.org/10.1088/2515-7620/abb0db.

Carvalho, R., L. Rausch, H.K. Gibbs, M.G. Bastos Lima, P. Bernasconi, A.P. Valdiones, A. Vasconcelos, and V. Silgueiro. 2024. Illegal deforestation in Mato Grosso: How Loopholes in implementing Brazil's forest code endanger the soy sector. *Land* 13: 1828. https://doi.org/10.3390/land13111828.

Carvalho, W.D., K. Mustin, R.R. Hilário, I.M. Vasconcelos, V. Eilers, and P.M. Fearnside. 2019. Deforestation control in the Brazilian Amazon: A conservation struggle being lost as agreements and regulations are subverted and bypassed. *Perspectives in Ecology and Conservation* 17: 122–130. https://doi.org/10.1016/j.pecon.2019.06.002.

Chen, B.K., B. Seligman, J.W. Farquhar, and J. Goldhaber-Fiebert. 2011. Multi-Country analysis of palm oil consumption and cardiovascular disease mortality for countries at different stages of economic development: 1980–1997. *Global Health* 7: 45. https://doi.org/10.1186/1744-8603-7-45.

Colchester, M., and S. Chai, eds. 2013. Conflict or Consent? The oil palm sector at a crossroads, FPP, Sawit Watch and TUK INDONESIA. http://www.forestpeoples.org/sites/fpp/files/publication/2013/11/conflict-or-consentenglishlowres.pdf.

Cramb, R., and J.F. McCarthy. 2016. Characterising oil palm production in Indonesia and Malaysia. In *The oil palm complex*, ed. R. Cramb and J.F. McCarthy. Singapore, NUS Press.

Davis, S.J., K. Alexander, J. Moreno-Cruz, et al. 2024. *Food without Agriculture. Nat Sustain* 7: 90–95. https://doi.org/10.1038/s41893-023-01241-2.

Dinis-Oliveira, R.J., J.A. Duarte, A. Sánchez-Navarro, F. Remião, M.L. Bastos, and F. Carvalho. 2008. Paraquat poisonings: Mechanisms of lung toxicity, clinical features, and treatment. *Critical Reviews in Toxicology* 38 (1): 13–71. https://doi.org/10.1080/10408440701669959.

Dreoni, I., Z. Matthews, and M. Schaafsma. 2022. The impacts of soy production on multi-dimensional well-being and ecosystem services: A systematic review. *Journal of Cleaner Production* 335: 130182. https://doi.org/10.1016/j.jclepro.2021.130182.

EA. 2019. Guidance: Site-specific quality numeric permit limits: discharges to surface water and groundwater. Environment Agency. https://www.gov.uk/government/.

EPA. 1993. R.E.D facts: Glyphosate EPA-738-F-93-011. US Environmental Protection Agency. https://www3.epa.gov/pesticides/chem_search/reg_actions/reregistration/fs_PC-417300_1-Sep-93.pdf.

FAS. 2023. Production, supply and distribution database. US Department of Agriculture Foreign Agricultural Service. https://apps.fas.usda.gov/psdonline/.

FAS. 2024a. Production—Palm Oil. US Department of Agriculture Foreign Agricultural Service. https://fas.usda.gov/data/production/commodity/4243000.

FAS. 2024b. Production—Soybeans. US Department of Agriculture Foreign Agricultural Service. https://fas.usda.gov/data/production/commodity/2222000.

Fearnside, P.M. 2017. Deforestation of the Brazilian Amazon. In *Oxford research encyclopedia of environmental science*, ed. H. Shugart, 1–53. New York: Oxford University Press. https://doi.org/10.1093/acrefore/9780199389414.013.102.

Fehlenberg, V., M. Baumann, N.I. Gasparri, M. Piquer-Rodríguez, G. Gavier-Pizarro, and T. Kuemmerle. 2017. The role of soybean production as an underlying driver of deforestation in the South American Chaco. *Global Environmental Change* 45: 24–34. https://doi.org/10.1016/j.gloenvcha.2017.05.001.

Fleiss, S., Hill, J.K. McClean, C. and Lucey, J.M. 2018. Potential Impacts of Climate CHange on Oil Palm Cultivation, Sensor Project. http://www.sensorproject.net/wp-content/uploads/2018/01/Climate-change-report-FINAL.pdf.

Foster, W.A., J.L. Snaddon, E.C. Turner, T.M. Fayle, T.D. Cockerill, M.D.F. Ellwood, G.R. Broad, A.Y.C. Chung, P. Eggleton, C.V. Khen, and K.M. Yusah. 2011. Establishing the evidence base for maintaining biodiversity and ecosystem function in the oil palm landscapes of Southeast Asia. *Philosophical Transactions of the Royal Society B* 366: 3277–3291. https://doi.org/10.1098/rstb.2011.0041.

Fraanje, W., and T. Garnett. 2020. *Soy: food, feed, and land use change. (Foodsource: Building Blocks).* Food Climate Research Network, University of Oxford.

George, J., S.K. Thakur, R.C. Tripathi, L.C. Ram, A. Gupta, and S. Prasad. 2010. Impact of coal industries on the quality of Damodar river water. *Toxicological & Environmental Chemistry* 92 (9): 1649–1664. https://doi.org/10.1080/02772241003783737.

GFA. 2024. Palm oil, Yale school of the environment global forest atlas. https://globalforestatlas.yale.edu/land-use/industrial-agriculture/palm-oil.

Gordon, A., and J. Jackson. 2017. Case study: application of appropriate technologies to improve the quality and safety of coconut water. In *Food safety and quality systems in developing countries: Volume Two: Case studies of effective implementation*, ed. A. Gordon. Academic Press. https://doi.org/10.1016/C2013-0-19360-9.

GWIS. 2024. GWIS annual statistics for Brazil, Global Wildfire Information Service. https://gwis.jrc.ec.europa.eu/apps/gwis.statistics/estimates/BRA/2024.

Guereña, A. 2013. The Soy Mirage, Oxfam Research Reports. https://www.oxfam.org/en/research/soy-mirage.

Hassan, M.A., S. Yacob, Y. Shirai, and Y. Hung. 2005. Treatment of palm oil wastewaters. *Waste Treat. Food Processing Industry* 2005: 101–117. https://doi.org/10.1201/9781420037128.ch4.

Henders, S., U.M. Persson, and T. Kastner. 2015. Trading forests. *Environmental Research Letters* 10: 125012. https://doi.org/10.1088/1748-9326/10/12/125012.

IndexMundi. 2024. Palm Oil production by country in 1000MT, IndexMundi. https://www.indexmundi.com/agriculture/?commodity=palm-oil.

IPAD. 2024. Brazil: Soybean production. International Production Assessment Division, USDA Foreign Agricultural Service. https://ipad.fas.usda.gov/countrysummary/Default.aspx?id=BR&crop=Soybean.

IPCC. 2019. Climate change and land An IPCC Special Report on climate change, desertification, land degradation, sustainable land management, food security, and greenhouse gas fluxes in terrestrial ecosystems, InterGovernmental Panel on Climate Change. https://www.ipcc.ch/site/assets/uploads/2019/11/SRCCL-Full-Report-Compiled-191128.pdf.

Lee, J.S.H., J. Ghazoul, K. Obidzinski, et al. 2014. Oil palm smallholder yields and incomes constrained by harvesting practices and type of smallholder management in Indonesia. *Agronomy for Sustainable Development* 34: 501–513. https://doi.org/10.1007/s13593-013-0159-4.

Le Page, M., 2023, Soya beans made more meat-like by adding genes for pig proteins, NewScientist, 28 June 2023, https://www.newscientist.com/article/2380087-soya-beans-made-more-meat-like-by-adding-genes-for-pigproteins/.

Li, D., and S. Liu. 2019. Detection of river water quality. In *Water quality monitoring and management: Basis, technology and case studies*, ed. D. Li and S. Liu. London: Academic Press.

Li, T.M. 2015. Social impacts of oil palm in Indonesia: A gendered perspective from West Kalimantan, CIFOR Occasional Paper No. 124, Bogor, Indonesia, Center for International Forestry Research. https://doi.org/10.17528/cifor/005579.

Loganathan, L., K.M. Subramaniam, A.K. Radhakrishnan, Y.-M. Choo, and K.-T. Teng. 2017. Health-promoting effects of red palm oil: Evidence from animal and human studies. *Nutrition Reviews* 75 (2): 98–113. https://doi.org/10.1093/nutrit/nuw054.

Lokman, N.A., A.M. Ithnin, W.J. Yahya, and M.A. Yuzir. A brief review on biochemical oxygen demand (BOD) treatment methods for palm oil mill effluents (POME). *Environmental Technology & Innovation* 21: 101258. https://doi.org/10.1016/j.eti.2020.101258.

Maluin, F.N., M.Z. Hussein, and A.S. Idris. 2020. An overview of the oil palm industry: Challenges and some emerging opportunities for nanotechnology development. *Agronomy* 10: 356. https://doi.org/10.3390/agronomy10030356.

McBride, K. 2024. Malaysia: Oilseeds and products update, GAIN Report number ID2024-0007, Global Agricultural Information Network, USDA Foreign Agricultural Service. https://apps.fas.usda.gov/newgainapi/api/Report/DownloadReportByFileName?fileName=Oilseeds%20and%20Products%20Update_Kuala%20Lumpur_Malaysia_MY2024-0007.pdf.

McCarthy, J.F., and R. Cramb. 2009. Policy narratives, landholder engagement, and oil palm expansion on the Malaysian and Indonesian frontiers. *The Geographical Journal* 175 (2): 112–123. https://doi.org/10.1111/j.1475-4959.2009.00322.x.

Mekhilefa, S., S. Sigaa, and R. Saidurb. 2011. A review on palm oil biodiesel as a source of renewable fuel. *Renewable and Sustainable Energy Reviews* 15: 1937–1949. https://doi.org/10.1016/j.rser.2010.12.012.

Mosaic. 2024. The silver lining soybean strategy. Mosaic. https://www.cropnutrition.com/resource-library/the-silver-lining-soybean-strategy/.

MPOC. 2016. Q & A Series On Palm Oil—Facts On Malaysian Palm Oil, Malaysian Palm oil Council. https://www.mpoc.org.my/q-a-series-on-palm-oil-facts-on-malaysian-palm-oil/.

Murphy, D.J., K. Goggin, and R.R.M. Paterson. 2021. Oil palm in the 2020s and beyond: Challenges and solutions. *CABI Agric Biosci* 2: 39. https://doi.org/10.1186/s43170-021-00058-3.

Nepstad, D., D. McGrath, C. Stickler, A. Alencar, A. Azevedo, B. Swette, T. Bezerra, M. DiGiano, J. Shimada, R. Seroa da Motta, E. Armijo, L. Castello, P. Brando, M.C. Hansen, M. McGrath-Horn, O. Carvalho, and L. Hess. 2014. Slowing Amazon deforestation through public policy and interventions in beef and soy supply chains Science, New Series 344 (6188): 1118–1123.

Okereke, J.N., and R.C. Ginikanwa. 2020. Environmental impact of palm oil mill effluent and its management through biotechnological approaches. *International Journal of Advanced Research in Biological Sciences* 7 (7): 117–127. https://doi.org/10.22192/ijarbs.2020.07.07.014.

Oliveira, G., and S. Hecht. 2016. Sacred groves, sacrifice zones and soy production: Globalization, intensification and neo-nature in South America. *The*

Journal of Peasant Studies 43 (2): 251–285. https://doi.org/10.1080/03066150.2016.1146705.
ONG Reportér Brasil. 2008. Brazil of biofuels: Impacts of crops on land, environment and society: Soy Castor Bean, Biofuel Watch Center, Reportér Brasil. https://reporterbrasil.org.br/wp-content/uploads/2024/03/brazil_of_biofuels_v1.pdf.
Pichler, M. 2013. "People, Planet & Profit": Consumer-oriented hegemony and power relations in palm oil and agrofuel certification. *Journal of Environment & Development* 22 (4): 370–390.
Piras, S., V.J. Wesz Jr., and S. Ghinoi. 2021. Soy Expansion, Environment, and Human Development: An Analysis across Brazilian Municipalities. *Sustainability.* 13 (13): 7246. https://doi.org/10.3390/su13137246.
Poku, K. 2002. Small-Scale Palm Oil Processing in Africa, FAO Agricultural Services Bulletin 148, Rome: Food and Agriculture Organization of the United Nations. https://www.fao.org/4/y4355e/y4355e00.htm.
Pye, O. 2019. Commodifying sustainability: Development, nature and politics in the palm oil industry. *World Development* 121: 218–228. https://doi.org/10.1016/j.worlddev.2018.02.014.
Qi, D., J. Zhu, and X. Wang. 2023. Nitrogen loss via runoff and leaching from paddy fields with the proportion of controlled-release urea and conventional urea rates under alternate wetting and drying irrigation. *Environmental Science and Pollution Research* 30: 61741–61752. https://doi.org/10.1007/s11356-023-26480-w.
Qiu, Y., Y. Zu, C. Song, M. Xie, Y. Qi, Y. Kansha, and Y. Kitamura. 2019. Soybean processing wastewater purification via Chlorella L166 and L38 with potential value-added ingredients production. *Bioresource Technology Reports* 7: 100195. https://doi.org/10.1016/j.biteb.2019.100195.
Rahmanulloh, A. 2024. Indonesia: Oilseeds and products update, GAIN Report number ID2024-0048, Global Agricultural Information Network, USDA Foreign Agricultural Service. https://apps.fas.usda.gov/newgainapi/api/Report/DownloadReportByFileName?fileName=Oilseeds%20and%20Products%20Update_Jakarta_Indonesia_ID2024-0048.pdf.
Reddington, C.L., L. Conibear, S. Robinson, C. Knote, S.R. Arnold, and D.V. Spracklen. 2021. Air pollution from forest and vegetation fires in Southeast Asia disproportionately impacts the poor. GeoHealth 5 (9). https://doi.org/10.1029/2021GH000418.
Renaldi, A., 2022, ndonesia's giant capital city is sinking. Can the government's plan save it?, National Geographic, 29 July 2022, https://www.nationalgeographic.com/environment/article/indonesias-giant-capital-city-is-sinking-can the-governments-plan-save-it

Rist, L., L. Feintrenie, and P. Levang. 2010. The livelihood impacts of oil palm: Smallholders in Indonesia. *Biodiversity and Conservation* 19: 1009–1024. https://doi.org/10.1007/s10531-010-9815-z.

RSPO. 2015. A shared vision. 100% sustainable palm oil in Europe: A snapshot of national initiatives. Geneva: Roundtable on Sustainable Palm Oil.

RSPO. 2024. As an organisation: Membership categories. https://rspo.org/as-an-organisation/membership/membership-categories/.

Sheil, D., A. Casson, E. Meijaard, M. van Noordwijk, J. Gaskell, J. Sunderland-Groves, K. Wertz, and M. Kanninen. 2009. The impacts and opportunities of oil palm in Southeast Asia: What do we know and what do we need to know? CIFOR Occasional Paper No. 51, Bogor, Indonesia, Center for International Forestry Research.

Sherman, J., M. Voigt, M. Ancrenaz, S.A. Wich, I.N. Qomariah, E. Lyman, E. Massingham, and E. Meijaard. 2022. Orangutan killing and trade in Indonesia: Wildlife crime, enforcement, and deterrence patterns. *Biological Conservation* 276: 109744. https://doi.org/10.1016/j.biocon.2022.109744.

Sneddon, S.T. 2021. Corruption, organised crime and species extinction in South East Asia. *The Resolution Journal* 2. https://www.jerseylawcommission.org.je/_files/ugd/f5ec37_c7fc0e0c6e2548d2a8068ee2f4d4bcc1.pdf.

Statista. 2020. Vegetable oil yields per hectare worldwide as of 2020, by crop plant (in metric tons per hectare). Statista. https://www.statista.com/statistics/1263201/vegetable-oil-yields-per-hectare-worldwide-by-type-globally/.

Stone, E.A., J.J. Schauer, B.B. Pradhan, P.M. Dangol, G. Habib, C. Venkataraman, and V. Ramanathan. 2010. Characterization of emissions from South Asian biofuels and application to source apportionment of carbonaceous aerosol in the Himalayas. *Journal of Geophysical Research. Atmospheres* 115 (6). https://doi.org/10.1029/2009JD011881.

Su Hongyang, S., Z. Yalei, Z. Chunmin, Z. Xuefei, and L. Li Jinpeng. 2011. Cultivation of Chlorella pyrenoidosa in soybean processing wastewater. *Bioresource Technology* 102 (21): 9884–9890. https://doi.org/10.1016/j.biortech.2011.08.016.

Sun, Y., N. Neelakantan, Y. Wu, R. Lote-Oke, A. Pan, and R.M. van Dam. 2015. Palm oil consumption increases LDL cholesterol compared with vegetable oils low in saturated fat in a meta-analysis of clinical trials 1, 2, 3. *The Journal of Nutrition* 145 (7): 1549–1558. https://doi.org/10.3945/jn.115.210575.

Tomasz OE, Viliker AS, Rondinone G. 2024. The economic cost of extreme and severe droughts in soybean production in Argentina, Contaduría y Administracíon 64 (1): 86. http://www.cya.unam.mx/index.php/cya/article/view/1422.

TrendEconomy. 2024. World merchandise exports and imports by commodity (HS): Soya beans (soybeans): Imports and exports 2023. TrendEconomy. https://trendeconomy.com/data/commodity_h2/1201.

USFDA. 2024. GMO Crops, Animal Food, and Beyond, United States Food and Drug Administration, https://www.fda.gov/food/agricultural-biotechnology/gmo-crops-animal-food-and-beyond.

Voora, V., Larrea, C., Bermudez, S., Balino, S. 2020. *Global Market Report: Palm Oil, International Institute for Sustainable Development/ State of Sustainability Initiatives*. https://www.iisd.org/system/files/publications/ssi-globalmarket-report-palm-oil.pdf

Vijay, V., S.L. Pimm, N.C. Jenkins, and S.J. Smith. 2016. The impacts of oil palm on recent deforestation and biodiversity loss. PLoS ONE 11 (7): e0159668. https://doi.org/10.1371/journal.pone.0159668.

WBA. 2024. Armaggi Group. World Benchmarking Alliance. https://www.worldbenchmarkingalliance.org/publication/food-agriculture/companies/amaggi-group-3/.

Weinhold, W., E. Killick, and E.J. Reis. 2013. Soybeans, Poverty and Inequality in the Brazilian Amazon. *World Development*. 52: 132–143, https://doi.org/10.1016/j.worlddev.2012.11.016.

White R. 2013. Environmental Harm: An Eco-lustice perspective, Policy Press.

Wu, T.Y., A.W. Mohammad, J.M. Jahim, and N. Anuar. 2010. Pollution control technologies for the treatment of palm oil mill effluent (POME) through end-of-pipe processes. *Journal of Environmental Management* 91 (7): 1467–1490. https://doi.org/10.1016/j.jenvman.2010.02.008.

WWF-UK. 2024. 8 things to know about palm oil. https://www.wwf.org.uk/updates/8-things-know-about-palm-oil.

Yacob, S., M.A. Hassan, Y. Shirai, M. Wakisaka, and S. Subash. 2006. Baseline study of methane emission from anaerobic ponds of palm oil mill effluent treatment. *Science of the Total Environment* 366: 187–196.

Yu, H.-Q., Z.-H. Hu, Z.-Q. Hong, and G.W. Gu. 2002. Performance of an anaerobic filter treating soybean processing wastewater with and without effluent recycle. *Process Biochemistry* 38: 507–513.

Yusoff, S. 2006. Renewable energy from palm oil—innovation on effective utilization of waste. *Journal of Cleaner Production* 14: 87–93. https://doi.org/10.1016/j.jclepro.2004.07.005.

Zaied, B.K., M. Nasrullah, Md.N.I. Siddique, A.W. Zularisam, L. Singh, and S. Krishnan. 2020. Co-digestion of palm oil mill effluent for enhanced biogas production in a solar assisted bioreactor: Supplementation with ammonium bicarbonate. *Science of The Total Environment* 706: 136095. https://doi.org/10.1016/j.scitotenv.2019.136095.

Zaied, B.K., Md.N.I. Siddique, M. Nasrullah, L. Singh, Z.B.A. Wahid, and M.F. Ahmad. 2019. Application of solar assisted bioreactor for biogas production

from palm oil mill effluent co-digested with cattle manure. *Environmental Technology & Innovation* 16: 100446. https://doi.org/10.1016/j.eti.2019.100446.
Zhang, Y and L. Chen. 2021. Soy Economics in Mato Grosso. https://storymaps.arcgis.com/stories/830534e420864582bc1b9f3c2ef19e93.
Ziegert, R.M., and M. Sotirov. 2024. Regulatory politics and hybrid governance: The case of Brazil's Amazon Soy Moratorium. *Global Environmental Change* 88: 102916. https://doi.org/10.1016/j.gloenvcha.2024.102916.

CHAPTER 6

Conclusion

Abstract This concluding chapter reflects on the book's exploration of eco-justice in industrial food systems, aligning its analysis with green criminology. It highlights how environmental harm, regulatory failures, and corporate exploitation drive injustices across food production, with many legally sanctioned practices being morally indefensible. The chapter revisits key themes, including environmental, ecological, and species justice, underscoring how deforestation, pollution, and habitat destruction disproportionately affect marginalised communities and non-human species. It also critiques the entrenched economic structures that resist change due to corporate profit motives. Additionally, the chapter applies Routine Activity Theory to food system injustices, arguing that financial incentives motivate corporations, vulnerable populations serve as suitable targets, and weak regulations allow exploitation to continue unchecked. The chapter advocates for sustainable agriculture, equitable food systems, corporate accountability, and interdisciplinary collaboration as essential pathways for reform. Ultimately, it emphasises that meaningful change requires consumer awareness and systemic transformation, challenging readers to reconsider their role in shaping the future of food production.

Keywords Green criminology · Environmental justice · Corporate accountability · Sustainable food systems · Routine activity theory

© The Author(s), under exclusive license to Springer Nature Switzerland AG 2025

S. Sneddon, *Food, Environmental Degradation and Injustices*, Palgrave Studies in Green Criminology,
https://doi.org/10.1007/978-3-031-87900-5_6

The book's exploration of eco-justice themes as they apply to industrial food production systems aligns closely with green criminology, a subfield dedicated to examining environmental harm, regulatory failures, and the role of state and corporate actors in perpetuating ecological injustice (White 2021). Green criminology's focus on both "illegal" and "legal" harms underscores the pervasive nature of exploitation in the global food industry. Across the chapters, it is evident that many harms—such as deforestation for palm oil and soya, overfishing, and the degradation of ecosystems—may be legally sanctioned yet morally indefensible. These practices often benefit powerful stakeholders while disproportionately impacting marginalised communities, Indigenous peoples, and non-human species.

The massive commercialisation of the global industrial food complex means that there is going to be considerable resistance to any change which has the potential to reduce corporate profits. As Mariam Stamp Dawkins puts it in relation to animals, if there is a call that "animals should be kept in conditions that give less human food or make it more expensive, then the case will have to be particularly clear and watertight, because every excuse will be found to turn it down" (Dawkins 2016, p. 4). What is true for beef, pigs, salmon, and prawns is equally true for potatoes, tomatoes, palm oil, and soya. Indeed, one theme emerging from all of the substantive chapters is how embedded the economic argument for the *status quo* has become over the past few decades.

The book has focused on a number of key themes in each chapter:

Environmental Justice

The unequal distribution of environmental risks and benefits is a recurring theme. From meat production's reliance on concentrated animal feeding operations (CAFOs) to aquaculture's ecological impact, the environmental burdens are disproportionately borne by marginalised and Indigenous communities. This aligns with White's framework, which emphasises the intersection of social inequity and ecological harm. For instance, the health risks posed by CAFOs—such as respiratory illnesses and zoonotic disease outbreaks—primarily affect low-income and minority populations living near these facilities (Chap. 2). Migrant workers are also disproportionately affected by environmental injustices, as seen in the dangerous conditions faced by workers on European tomato farms (Chap. 3), Thai fisheries (Chap. 4), and Malaysian oil palm plantations (Chap. 5). For example, in Italy and Spain, undocumented migrant workers on tomato farms endure long hours, substandard wages, and exposure to pesticides without adequate protection, while in Thailand's fishing industry, labourers frequently encounter abuse and unsafe working conditions. Similarly, oil palm plantations in Malaysia rely heavily on migrant labour, often under precarious contracts that provide minimal protection against workplace hazards.

Ecological Justice

The chapters illustrate the profound impacts of food production on ecosystems and biodiversity. The destruction of habitats for cattle ranches, palm oil, and soya cultivation exemplifies ecological injustice, as entire ecosystems are sacrificed for short-term economic gains. For example, Chap. 5 details how the deforestation in Indonesia and Malaysia for palm oil plantations contributes to biodiversity loss, threatening species such as orangutans and Sunda tigers. Similarly, aquaculture's overuse of antibiotics and chemical inputs creates hypoxic zones in aquatic environments, decimating marine biodiversity (Chap. 4). Even tomato and potato farming (Chap. 3) relies heavily on monoculture practices, exacerbating soil degradation and threatening pollinator populations.

Species Justice:

The ethical treatment of non-human species emerges as a critical issue. Industrial farming practices, whether for cattle, fish, or prawns, often prioritise efficiency and profit over the welfare of animals. The commodification of living beings raises significant ethical concerns, aligning with criminological debates on speciesism and the moral obligations humans owe to other species. For instance, Chap. 2 highlights the suffering of livestock in CAFOs, where animals are subjected to overcrowding, poor sanitation, and inhumane slaughter methods to maximise yield. Similarly, Chap. 4 examines the high-density confinement of fish in aquaculture operations, where limited space and unsanitary conditions often result in stress-induced diseases and high mortality rates. In both cases, animals are treated as mere commodities, with their welfare consistently subordinated to the demands of industrial-scale production. This exploitation reflects broader systemic issues, calling for stronger regulatory frameworks and ethical interventions to protect the rights and well-being of non-human species.

Global Inequities

Global food systems are characterised by stark inequalities. The economic benefits of agricultural and aquaculture industries disproportionately accrue to corporations and wealthy consumers, while environmental and social costs are externalised to vulnerable populations. For example, Chap. 3 highlights the exploitation of migrant labour in tomato and potato farming, where workers face substandard wages and hazardous conditions. Chapter 5 further illustrates how the profits from palm oil and soya industries are concentrated among multinational corporations, while Indigenous communities bear the brunt of deforestation and pollution.

Corruption and Regulatory Failures

Corruption and weak regulatory frameworks enable environmental degradation and labour exploitation. As discussed in Chap. 5, the palm oil industry's rapid expansion has been facilitated by collusion between corporations, corrupt officials, and criminal networks (see also Sneddon 2021). For example, the allocation of land for oil palm plantations often disregards the rights of Indigenous communities, displacing them without adequate compensation. Additionally, voluntary initiatives like the Roundtable on Sustainable Palm Oil (RSPO) have faced criticism for their lack of enforcement mechanisms, allowing certified companies to continue harmful practices. In the aquaculture sector (Chap. 4), regulatory gaps enable the overuse of antibiotics and chemicals, contributing to environmental damage and antibiotic resistance. Similarly, in the meat industry (Chap. 2), inadequate oversight of CAFOs exacerbates pollution and health risks for nearby marginalised communities. Addressing these systemic failures requires a robust legal and regulatory response, including stricter enforcement, transparency measures, and international cooperation to hold stakeholders accountable.

Though these themes have been considered in some depth in the chapters, what has not really been addressed yet is the rationale for this behaviour. One possible approach is to look back on the preceding chapters through the lens of Cohen and Felson's *Routine Activity Theory*. (Cohen and Felson 1979). This was designed as a means of trying to explain what the factors could be that facilitate criminal activities within everyday life. Although much of the behaviours outlined in this book falls under the "legal harms" umbrella, some of it (e.g., using victims of trafficking as workers (Chap. 3)) is definitely an "illegal harm." As such, it is worth considering how well these acts fit within the model. Cohen and Felson suggest (1979, p. 589) that three elements need to converge in order to influence crime:

1. Motivated Offenders;
2. Suitable Targets; and
3. The absence of capable guardians against a violation.

If we initially focus on the motivated offender category, it becomes apparent though all four chapters that there are individuals and corporations who possess the intention and ability to carry out the acts. Largely, to revisit Dawkins (2016), this motivation is financial. In Chap. 2, we saw that there were significant economic gains to be made due to the economies of scale of housing pigs and cattle in Concentrated Animal Feed Operations (CAFOs), and that the practice is becoming more widespread, despite the problems of pollution and animal welfare which are associated with the practice. The same was true for the deforestation of rainforests (whether the Amazon or elsewhere) to create grazing pasture for cattle. Chapter 3 revealed the explosion in the area under glass in the "controlled-environment agriculture" areas of Almería, where tomato growers have prioritised maximisation of income over everything else. Also in Chap. 3, the role of Organised Criminal Groups and the *caporale* in exploiting both migrant workers and trafficking victims were noted, along with controversies with the role of the Chinese State and the Uighur re-education camps. Chapter 4 showed us that aquaculture companies are now responsible for over half of global seafood production, and that overly high stocking densities and overuse of chemicals and antibiotics are leading to significant welfare issues. In addition, there are associated issues of habitat loss, especially in coastal mangrove areas. Chapter 5 led us into the forests of Southeast Asia, and to the massive deforestation which has occurred, and still occurs, as areas are cleared for palm oil production. We also saw reports of flagrant labour law violations in Indonesia especially, and abuse of migrant workers in Malaysia. The Soy industry too is responsible for considerable deforestation, and, despite the Amazon Soy Moratorium, we also saw the lengths to which growers were prepared to go to evade such restrictions.

The corporations and individuals who are engaged in these practices are not evil, per se, indeed as the Lord Chancellor Baron Edward Thurlow (is alleged to have) said "Did you ever expect a corporation to have a conscience, when it has no soul to be damned, and no body to be kicked?" (Coffee 1981) or "Corporations have neither bodies to be punished, nor souls to be condemned; they therefore do as they like" (Poynder 1844, p. 268). Corporations are amoral, and their motivation to commit these acts is triggered by economics.

The second element of the Routine Activity Theory is the need for a suitable target for the harm. For "illegal harms," other models have emerged to develop Cohen and Felson's original, including Clarke's (1999) CRAVED model. CRAVED stands for Concealable, Removable, Available, Valuable, Enjoyable, and Disposable, and these are the elements which Clarke posited were the characteristics that make an item attractive to thieves (Clarke 1999). However, although cattle theft is a significant issue (see, for example, Marigat 2023) most of the harms identified in the book involve damage *to* things, rather than the taking *of* things. There are potentially three categories of "suitable target" identified in the book, each of which will be dealt with in turn. Firstly, natural resources. Ecosystems like rainforests, marine environments, and arable lands are exploited for agricultural and aquaculture production. These are prime targets due to their economic value and relative lack of direct protection in some regions. Rainforests are targeted directly by cattle ranchers in Brazil (Chap. 2), sometimes with the tacit or explicit support of the government (see, Roy 2022) and indirectly by Soy farmers (Chap. 5) attempting to circumvent the ASM. Marine environments are compromised directly by aquaculture (Chap. 4) and through the pollution run-off from the use of pesticides, herbicides, fungicides, and antibiotics in the farming of cattle, pigs, tomatoes, potatoes, fish, prawns, palm oil, and soy. Arable land too is compromised, either by its conversion of "temporal crop diversification" which covers alternating species of crops, using cover crops and a more "patchwork" landscape of crops (Guinet et al. 2023) with large monoculture plantation (Chap. 3 for tomatoes and potatoes, but especially Chap. 5 for soy farming in Mato Grosso, Brazil).

The second category of suitable target is the labour force. Vulnerable populations, such as migrant workers or Indigenous communities, are exploited as they often lack the social or legal means to resist unfair treatment. In Chap. 2, we saw how rural pig farmers in China were being dispossessed, and that small local farmers were unable to compete with the larger industrial entities. We also saw that farmers in Italy were being intimidated by OCGs *Cosa Nostra* and *Sacra Corona Unita* who were laundering cattle subsidies. Chapter 3 looked at the allegations of forced labour in China's Xinjiang region, and the exploitation of undocumented migrants in Spain, both linked to the tomato industry. Exploitation of workers is also an issue in aquaculture, as demonstrated in Chap. 4. Whether these were prisoners in Colorado's tilapia sector, migrant workers in Southeast Asian prawn farms, or inadequate safety provisions in Chilean salmon farms, this group of people is definitely being targeted by these industries, primarily as they are cheaper to "employ."

The final category of potential target for harm is consumers. This is a less obvious target, but an important one, nonetheless. Consumers may be misled into supporting corporations engaged in legal or illegal harms. There are countless organisations offering consumers comfort about the standards to which their food was produced—whether that is the Roundtable on Sustainable Palm Oil production (Chap. 4), the National Farmers' Union *Red Tractor* in the UK, Fair Trade, Rainforest Alliance, or any other. All of these organisations set limits and requirements which exceed any baseline required by national law, and the idea is that their certification becomes a good marker for the environmental, ethical, or other credentials of the company. When these work, they are excellent, but in Chap. 5 we saw that the RSPO faces criticism linking its members to deforestation and worker safety (Pye 2019) and about its fundamental structure (Pichler 2013). Fair Trade too has come in for its fair share of criticism over employment standards (Henderson 2008) though Henderson's criticisms were stridently opposed by Smith (2009). This muddled sector leaves the consumer in a position where they are ripe to be exploited.

The third element required for Routine Activity Theory is the absence of a capable guardian. This refers to lack of individuals or systems that could prevent the harm, whether these are formal, such as law enforcement or other official regulatory bodies, or more informal like consumers or vigilant community members. Again, this can be subdivided, this time into two areas. Firstly, weak or absent regulatory frameworks. In regions which lack robust environmental, labour, and oversight laws, or where the regulatory bodies are unwilling or unable to enforce existing regulations, an environment where exploitation can flourish is created. In Chap. 2, we saw that many countries lack sufficiently stringent rules to control pollution from CAFOs, leading to issues like water contamination from pollutant run-off causing eutrophication and anoxia in rivers (Mallin 2000). We have also seen how OCGs in Italy have managed to exploit regulatory weaknesses by laundering money through cattle farming. Chapter 3 covered the imbalance between large corporations and small farmers in terms of seed patents and GM crops, and what Odong (2024) called the inherent and unresolved conflicts between the trade restrictive practices of the Cartagena Protocol on Biosafety and the GATT/WTO rules. The chapter also covered the labour exploitation in the tomato industry, and the need for tighter regulations and better enforcement to prevent reoccurrence. In Chap. 4, we noted two significant areas of weak regulation—the use of antibiotics and chemicals, which Thornber et al. (2019) note that the use of antibiotics is unregulated in many regions, and the rampant exploitation of fish workers in Thailand, where weak labour protections and enforcement allow unscrupulous operators to exploit workers, including cases of modern slavery. In Chap. 5, we saw that in Indonesia and Malaysia, policies surrounding land use are often described as inadequate or skewed in favour of agribusiness interests. For example, Cramb and McCarthy (2016) highlight how narratives of "unproductive" local agriculture justify deforestation, and Pye (2019) attributes 90% of deforestation to private palm oil corporations, even those which are RSPO certified. The RSPO and the ASM are both noted to be voluntary initiatives with no legal enforcement. Although they have reduced deforestation in some cases, loopholes (e.g., cattle ranchers selling deforested land to soy farmers) undermine their effectiveness.

The second element of the absence of capable guardians is the very nature and opacity of global trading patterns. The dispersed nature of global supply chains makes monitoring and accountability difficult. A palm oil plantation in Indonesia, for instance, may supply a multinational corporation headquartered in a country with stricter laws, yet the corporation remains shielded from liability. In Chap. 3, however, we saw that Directive 2024/1760 on Corporate Sustainability and Due Diligence requires EU Member States to pass legislation ensuring that all companies over a particular size carry out due diligence checks "regarding actual and potential human rights adverse impacts and environmental adverse impacts, with respect to their own operations, the operations of their subsidiaries, and the operations carried out by their business partners in the chains of activities of those companies" (Article 1(1)(a)). While the geographical scope of the Directive is limited to EU Member States, if the national enacting legislation is effectively enforced, it should go some way to lessening this part of the problem.

Cohen and Felson's Routine Activity Theory does explain much of the activity that has been discussed in the preceding chapters. They conclude their paper by saying:

> Rather than assuming that predatory crime is simply an indicator of social breakdown, one might take it as a byproduct of freedom and prosperity as they manifest themselves in the routine activities of everyday life (1979, p. 605).

If we take the same approach, it would suggest that the harms caused by the industrialisation of agriculture are not an indication of global breakdown but are just by-products of globalisation. That is almost undoubtedly true, but it does not mean that we should accept it as an immovable standard.

Although the thread through the book was the immediate cessation of the commercial production of, and trade in these commodities as food, this was acknowledged at the start as a "thought experiment." Despite the COVID-19 pandemic and the announcement in the spring of 2024 that H5N1 Avian Influenza had been detected in cows and humans (and 200 other species of land-based and water-based mammal) in the United States (CDC 2024), the likelihood of a John Christopher-style pandemic which wipes out particular foodstuffs is low. Each of the chapters demonstrated that, despite the environmental, ecological, and species justice arguments being predominantly in favour or ending some of the trades, the economic embeddedness would make it is impossible to achieve without triggering significant global problems.

However, this was never a viable option. What the book has done is demonstrate that in relation to almost all of the foodstuffs considered, current practices are entirely unsustainable, perpetuating cycles of environmental harm, social inequity, and species exploitation. This disappointing realisation (that there is so much need for change, even outside the meat industry) is also an opportunity to consider how we can change the status quo. Any changes to the way we produce food must have four elements at their heart.

Sustainable Agriculture and Aquaculture

Both agriculture and aquaculture feature in the UN Sustainable Development Goals, either directly (SDG 14: Life below water; SDG 15: Life on Land) or indirectly (as contributing factors to almost all of the other SDGs). The SDGs are set to expire in 2030 and, whatever global mechanisms replace them, it is essential that innovative practices are embraced to ensure true sustainability. Innovations in sustainable practices, such as regenerative agriculture and integrated multi-trophic aquaculture, can mitigate environmental impacts while supporting food security. For instance, regenerative agriculture practices, as discussed in Chap. 3, focus on restoring soil health through crop rotation and minimal tillage, reducing dependency on chemical inputs. Integrated multi-trophic aquaculture (IMTA), highlighted in Chap. 4, combines different species such as fish and shellfish in a single system, where waste from one species serves as nutrients for another, thus minimising environmental pollution and improving ecosystem efficiency.

Equitable Food Systems

Policies must address the structural inequalities that marginalise smallholder farmers and Indigenous communities. This includes enforcing land tenure rights and supporting local food sovereignty. Chapter 5 discusses how Indigenous communities displaced by palm oil plantations in Indonesia face significant economic and cultural losses. Supporting these communities through legal protections and investments in community-led farming initiatives can enhance their resilience and ensure fair participation in food systems. The Principles of Environmental Justice, as set out in Chap. 1, include provisions for participatory justice (Principle 5, Principle 7, Principle 13) and embedding these within future policies would be one mechanism for ensuring structural inequalities are lessened, if not removed.

Corporate Accountability

Holding corporations accountable for environmental and social harms is essential. This can be achieved through stricter regulations (such as the Environmental Permitting Regulations in the UK), supply chain transparency (such as the Directive 2024/1760 on Corporate Sustainability and Due Diligence), and consumer advocacy. Chapter 2 illustrates how the unchecked growth of CAFOs has led to severe pollution, including groundwater contamination from manure run-off and air pollution from ammonia emissions. These practices have significant public health implications, especially for low-income communities living near these facilities, who suffer from respiratory illnesses and other chronic health conditions. Additionally, weak oversight allows corporations to externalise costs, leaving taxpayers to bear the burden of environmental clean-up and healthcare expenses. Chapter 5 highlights similar issues in the palm oil industry, where voluntary sustainability certifications often fail to prevent deforestation and labour exploitation. Strengthening corporate accountability requires mandatory environmental reporting, robust enforcement of labour rights, and incentives for sustainable practices. Empowering consumers through clear labelling and ethical certifications can also drive systemic change by shifting market demands.

Interdisciplinary Collaboration

I have a background in law, particularly environmental law. This book sits within criminology, specifically green criminology. Creating the book required engaging with literature from environmental science, philosophy, politics, and a range of other disciplines. Bridging disciplinary gaps is the best way to provide a holistic understanding of food-related injustices and inform effective interventions. For example, Chap. 4's analysis of aquaculture demonstrates the value of integrating scientific research on ecosystem health with ethical considerations for labour and species welfare. The chapter illustrates how the overuse of antibiotics in fish farming contributes to both environmental harm and human health risks, requiring interdisciplinary strategies to balance ecological and labour interests. Chapter 5 explores how collaborative frameworks can address enforcement gaps, particularly in the palm oil industry, where voluntary sustainability standards often fail due to insufficient oversight and accountability. By combining criminological insights with scientific expertise and ethical governance, interdisciplinary approaches ensure that interventions are both technically effective and socially just. Such collaborations can also inform global policymaking, encouraging cross-sectoral solutions to complex challenges in food production and sustainability.

The interconnectedness of food systems, environmental degradation, and social injustice requires a comprehensive and systemic approach to addressing these challenges. By integrating criminological theories with eco-justice principles, this book provides a starting point for understanding and addressing some of the systemic issues embedded in global food production. Ultimately, achieving justice for humans, ecosystems, and non-human species requires a collective commitment to equity, sustainability, and ethical responsibility. To paraphrase US President Kennedy, we need to do these things not because they are easy, but because they are hard. If nothing in our approach to food production changes, and the global population rises by almost one quarter by 2020 as predicted (UN 2024), then the results will be catastrophic.

Consumers, including you, have power to change the way food is produced, and what food is produced. Small changes, if replicated often enough, can make multinational corporations change their behaviour. The most significant difference we can make is to make *informed choices* about what we eat. I promised at the start of the book that it was not going to be a polemic against or in favour of particular decisions, and I stick by that promise.

I will leave you with a question—what are you going to eat tonight?

Sources

CDC. 2024. Avian Influenza (Bird Flu), US Center for Disease Control. https://www.cdc.gov/bird-flu/situation-summary/mammals.html.

Clarke, R.V. 1999. Hot products: understanding, anticipating, and reducing demand for stolen goods. Police Research Series Paper 112. https://popcenter.asu.edu/sites/default/files/hot_products_understanding_anticipating_and_reducing_demand_for_stolen_goods_clarke_1999.pdf.

Coffee, J. 1981. No soul to damn. No body to kick: An Unscandalized Inquiry into the Problem of Corporate Punishment 79 Mich L. R. 386.

Cohen, L.E., and M. Felson. 1979. Social change and crime rate trends: A routine activity approach. *American Sociological Review* 44 (August): 588–608.

Cramb, R., and J.F. McCarthy. 2016. Characterising oil palm production in Indonesia and Malaysia. In *The oil palm complex*, ed. by R. Cramb and J.F. McCarthy, eds. Singapore, NUS Press.

Dawkins, M.S. 2016. *Why animals matter: Animal consciousness, animal welfare and human wellbeing*. Oxford: OUP.

Guinet, M., G. Adeux, S. Cordeau, E. Courson, R. Nandillon, Y. Zhang, and N. Munier-Jolain. 2023. Fostering temporal crop diversification to reduce pesticide use. *Nature Communications* 14. Article number: 7416. https://doi.org/10.1038/s41467-023-43234-x.

Henderson, D. 2008. Fair trade is counterproductive and unfair. *Economic Affairs* 28 (3): 62–64. https://doi.org/10.1111/j.1468-0270.2008.00847.x.

Mallin, M.A. 2000. Impacts of industrial animal production on rivers and estuaries: Animal-waste lagoons and sprayfields near aquatic environments may significantly degrade water quality and endanger health. *American Scientist* 88 (1): 26–37. https://www.jstor.org/stable/27857960.

Marigat, S.K. 2023. Managing cattle rustling by enhancing police-community cooperation in the Karamoja Cluster: Lessons from Baringo, Kenya. *African Security Review* 32 (1). https://doi.org/10.1080/10246029.2022.2141129.

Odong, L.-M. 2024. Reconciling the incongruence between the Cartagena protocol on biosafety to the convention on biological diversity and the GATT/WTO rules. *International Environmental Agreements: Politics, Law, and Economics*. https://doi.org/10.1007/s10784-024-09649-7.

Pichler, M. 2013. "People, Planet & Profit": Consumer-oriented hegemony and power relations in palm oil and agrofuel certification. *Journal of Environment & Development* 22 (4): 370–390.

Poynder, J. 1844. *Literary extracts from English and other works; collected during half a century: together with some original matter*. London: John Hatchard & Son.

Pye, O. 2019. Commodifying sustainability: Development, nature, and politics in the palm oil industry. *World Development* 121: 218–228. https://doi.org/10.1016/j.worlddev.2018.02.014.

Roy, D. 2022. Deforestation of Brazil's Amazon has reached a record high. What's being done? Council on Foreign Relations, 24 August. https://www.cfr.org/in-brief/deforestation-brazils-amazon-has-reached-record-high-whats-being-done.

Smith, A.M. 2009. Evaluating the criticism of fair trade. *Economic Affairs* 29 (4): 2–36. https://doi.org/10.1111/j.1468-0270.2009.01944.x.

Sneddon, S.T. 2021. Corruption, organised crime and species extinction in South East Asia. *The Resolution Journal* 2. https://www.jerseylawcommission.org.je/_files/ugd/f5ec37_c7fc0e0c6e2548d2a8068ee2f4d4bcc1.pdf.

Thornber, K., D. Verner-Jeffreys, S. Hinchliffe, M.M. Rahman, D. Bass, and T.R. Tyler. 2019. Evaluating antimicrobial resistance in the global shrimp industry. *Reviews in Aquaculture* 12: 966–986.

UN. 2024. Global Issues: Population, United Nations. https://www.un.org/en/global-issues/population.

White, R. 2021. *Theorising green criminology: Selected essays*. London: Routledge.

Index

© The Author(s), under exclusive license to Springer Nature Switzerland AG 2025

S. Sneddon, *Food, Environmental Degradation and Injustices*, Palgrave Studies in Green Criminology,
https://doi.org/10.1007/978-3-031-87900-5

The manufacturer's authorised representative in the EU is Springer Nature Customer Service Centre GmbH, Europaplatz 3, 69115 Heidelberg, Germany. If you have any concerns regarding our products, please contact ProductSafety@springernature.com

Printed and bound by CPI Group (UK) Ltd, Croydon, CR0 4YY
30/09/2026
02225595-0013